Handbook of Dough Fermentations

FOOD SCIENCE AND TECHNOLOGY

A Series of Monographs, Textbooks, and Reference Books

1. Flavor Research: Principles and Techniques, *R. Teranishi, I. Hornstein, P. Issenberg, and E. L. Wick*
2. Principles of Enzymology for the Food Sciences, *John R. Whitaker*
3. Low-Temperature Preservation of Foods and Living Matter, *Owen R. Fennema, William D. Powrie, and Elmer H. Marth*
4. Principles of Food Science
   Part I: Food Chemistry, *edited by Owen R. Fennema*
   Part II: Physical Methods of Food Preservation, *Marcus Karel, Owen R. Fennema, and Daryl B. Lund*
5. Food Emulsions, *edited by Stig E. Friberg*
6. Nutritional and Safety Aspects of Food Processing, *edited by Steven R. Tannenbaum*
7. Flavor Research: Recent Advances, *edited by R. Teranishi, Robert A. Flath, and Hiroshi Sugisawa*
8. Computer-Aided Techniques in Food Technology, *edited by Israel Saguy*

9. Handbook of Tropical Foods, *edited by Harvey T. Chan*
10. Antimicrobials in Foods, *edited by Alfred Larry Branen and P. Michael Davidson*
11. Food Constituents and Food Residues: Their Chromatographic Determination, *edited by James F. Lawrence*
12. Aspartame: Physiology and Biochemistry, *edited by Lewis D. Stegink and L. J. Filer, Jr.*
13. Handbook of Vitamins: Nutritional, Biochemical, and Clinical Aspects, *edited by Lawrence J. Machlin*
14. Starch Conversion Technology, *edited by G. M. A. van Beynum and J. A. Roels*
15. Food Chemistry: Second Edition, Revised and Expanded, *edited by Owen R. Fennema*
16. Sensory Evaluation of Food: Statistical Methods and Procedures, *Michael O'Mahony*
17. Alternative Sweeteners, *edited by Lyn O'Brien Nabors and Robert C. Gelardi*
18. Citrus Fruits and Their Products: Analysis and Technology, *S. V. Ting and Russell L. Rouseff*
19. Engineering Properties of Foods, *edited by M. A. Rao and S. S. H. Rizvi*
20. Umami: A Basic Taste, *edited by Yojiro Kawamura and Morley R. Kare*
21. Food Biotechnology, *edited by Dietrich Knorr*
22. Food Texture: Instrumental and Sensory Measurement, *edited by Howard R. Moskowitz*
23. Seafoods and Fish Oils in Human Health and Disease, *John E. Kinsella*
24. Postharvest Physiology of Vegetables, *edited by J. Weichmann*
25. Handbook of Dietary Fiber: An Applied Approach, *Mark L. Dreher*
26. Food Toxicology, Parts A and B, *Jose M. Concon*
27. Modern Carbohydrate Chemistry, *Roger W. Binkley*
28. Trace Minerals in Foods, *edited by Kenneth T. Smith*
29. Protein Quality and the Effects of Processing, *edited by R. Dixon Phillips and John W. Finley*
30. Adulteration of Fruit Juice Beverages, *edited by Steven Nagy, John A. Attaway, and Martha E. Rhodes*
31. Foodborne Bacterial Pathogens, *edited by Michael P. Doyle*
32. Legumes: Chemistry, Technology, and Human Nutrition, *edited by Ruth H. Matthews*
33. Industrialization of Indigenous Fermented Foods, *edited by Keith H. Steinkraus*
34. International Food Regulation Handbook: Policy • Science • Law, *edited by Roger D. Middlekauff and Philippe Shubik*
35. Food Additives, *edited by A. Larry Branen, P. Michael Davidson, and Seppo Salminen*
36. Safety of Irradiated Foods, *J. F. Diehl*

37. Omega-3 Fatty Acids in Health and Disease, *edited by Robert S. Lees and Marcus Karel*
38. Food Emulsions: Second Edition, Revised and Expanded, *edited by Kåre Larsson and Stig E. Friberg*
39. Seafood: Effects of Technology on Nutrition, *George M. Pigott and Barbee W. Tucker*
40. Handbook of Vitamins: Second Edition, Revised and Expanded, *edited by Lawrence J. Machlin*
41. Handbook of Cereal Science and Technology, *Klaus J. Lorenz and Karel Kulp*
42. Food Processing Operations and Scale-Up, *Kenneth J. Valentas, Leon Levine, and J. Peter Clark*
43. Fish Quality Control by Computer Vision, *edited by L. F. Pau and R. Olafsson*
44. Volatile Compounds in Foods and Beverages, *edited by Henk Maarse*
45. Instrumental Methods for Quality Assurance in Foods, *edited by Daniel Y. C. Fung and Richard F. Matthews*
46. *Listeria*, Listeriosis, and Food Safety, *Elliot T. Ryser and Elmer H. Marth*
47. Acesulfame-K, *edited by D. G. Mayer and F. H. Kemper*
48. Alternative Sweeteners: Second Edition, Revised and Expanded, *edited by Lyn O'Brien Nabors and Robert C. Gelardi*
49. Food Extrusion Science and Technology, *edited by Jozef L. Kokini, Chi-Tang Ho, and Mukund V. Karwe*
50. Surimi Technology, *edited by Tyre C. Lanier and Chong M. Lee*
51. Handbook of Food Engineering, *edited by Dennis R. Heldman and Daryl B. Lund*
52. Food Analysis by HPLC, *edited by Leo M. L. Nollet*
53. Fatty Acids in Foods and Their Health Implications, *edited by Ching Kuang Chow*
54. *Clostridium botulinum*: Ecology and Control in Foods, *edited by Andreas H. W. Hauschild and Karen L. Dodds*
55. Cereals in Breadmaking: A Molecular Colloidal Approach, *Ann-Charlotte Eliasson and Kåre Larsson*
56. Low-Calorie Foods Handbook, *edited by Aaron M. Altschul*
57. Antimicrobials in Foods: Second Edition, Revised and Expanded, *edited by P. Michael Davidson and Alfred Larry Branen*
58. Lactic Acid Bacteria, *edited by Seppo Salminen and Atte von Wright*
59. Rice Science and Technology, *edited by Wayne E. Marshall and James I. Wadsworth*
60. Food Biosensor Analysis, *edited by Gabriele Wagner and George G. Guilbault*
61. Principles of Enzymology for the Food Sciences: Second Edition, *John R. Whitaker*
62. Carbohydrate Polyesters as Fat Substitutes, *edited by Casimir C. Akoh and Barry G. Swanson*
63. Engineering Properties of Foods: Second Edition, Revised and Expanded, *edited by M. A. Rao and S. S. H. Rizvi*

64. Handbook of Brewing, *edited by William A. Hardwick*
65. Analyzing Food for Nutrition Labeling and Hazardous Contaminants, *edited by Ike J. Jeon and William G. Ikins*
66. Ingredient Interactions: Effects on Food Quality, *edited by Anilkumar G. Gaonkar*
67. Food Polysaccharides and Their Applications, *edited by Alistair M. Stephen*
68. Safety of Irradiated Foods: Second Edition, Revised and Expanded, *J. F. Diehl*
69. Nutrition Labeling Handbook, *edited by Ralph Shapiro*
70. Handbook of Fruit Science and Technology: Production, Composition, Storage, and Processing, *edited by D. K. Salunkhe and S. S. Kadam*
71. Food Antioxidants: Technological, Toxicological, and Health Perspectives, *edited by D. L. Madhavi, S. S. Deshpande, and D. K. Salunkhe*
72. Freezing Effects on Food Quality, *edited by Lester E. Jeremiah*
73. Handbook of Indigenous Fermented Foods: Second Edition, Revised and Expanded, *edited by Keith H. Steinkraus*
74. Carbohydrates in Food, *edited by Ann-Charlotte Eliasson*
75. Baked Goods Freshness: Technology, Evaluation, and Inhibition of Staling, *edited by Ronald E. Hebeda and Henry F. Zobel*
76. Food Chemistry: Third Edition, *edited by Owen R. Fennema*
77. Handbook of Food Analysis: Volumes 1 and 2, *edited by Leo M. L. Nollet*
78. Computerized Control Systems in the Food Industry, *edited by Gauri S. Mittal*
79. Techniques for Analyzing Food Aroma, *edited by Ray Marsili*
80. Food Proteins and Their Applications, *edited by Srinivasan Damodaran and Alain Paraf*
81. Food Emulsions: Third Edition, Revised and Expanded, *edited by Stig E. Friberg and Kåre Larsson*
82. Nonthermal Preservation of Foods, *Gustavo V. Barbosa-Cánovas, Usha R. Pothakamury, Enrique Palou, and Barry G. Swanson*
83. Milk and Dairy Product Technology, *Edgar Spreer*
84. Applied Dairy Microbiology, *edited by Elmer H. Marth and James L. Steele*
85. Lactic Acid Bacteria: Microbiology and Functional Aspects: Second Edition, Revised and Expanded, *edited by Seppo Salminen and Atte von Wright*
86. Handbook of Vegetable Science and Technology: Production, Composition, Storage, and Processing, *edited by D. K. Salunkhe and S. S. Kadam*
87. Polysaccharide Association Structures in Food, *edited by Reginald H. Walter*
88. Food Lipids: Chemistry, Nutrition, and Biotechnology, *edited by Casimir C. Akoh and David B. Min*
89. Spice Science and Technology, *Kenji Hirasa and Mitsuo Takemasa*

90. Dairy Technology: Principles of Milk Properties and Processes, *P. Walstra, T. J. Geurts, A. Noomen, A. Jellema, and M. A. J. S. van Boekel*
91. Coloring of Food, Drugs, and Cosmetics, *Gisbert Otterstätter*
92. *Listeria*, Listeriosis, and Food Safety: Second Edition, Revised and Expanded, *edited by Elliot T. Ryser and Elmer H. Marth*
93. Complex Carbohydrates in Foods, *edited by Susan Sungsoo Cho, Leon Prosky, and Mark Dreher*
94. Handbook of Food Preservation, *edited by M. Shafiur Rahman*
95. International Food Safety Handbook: Science, International Regulation, and Control, *edited by Kees van der Heijden, Maged Younes, Lawrence Fishbein, and Sanford Miller*
96. Fatty Acids in Foods and Their Health Implications: Second Edition, Revised and Expanded, *edited by Ching Kuang Chow*
97. Seafood Enzymes: Utilization and Influence on Postharvest Seafood Quality, *edited by Norman F. Haard and Benjamin K. Simpson*
98. Safe Handling of Foods, *edited by Jeffrey M. Farber and Ewen C. D. Todd*
99. Handbook of Cereal Science and Technology: Second Edition, Revised and Expanded, *edited by Karel Kulp and Joseph G. Ponte, Jr.*
100. Food Analysis by HPLC: Second Edition, Revised and Expanded, *edited by Leo M. L. Nollet*
101. Surimi and Surimi Seafood, *edited by Jae W. Park*
102. Drug Residues in Foods: Pharmacology, Food Safety, and Analysis, *Nickos A. Botsoglou and Dimitrios J. Fletouris*
103. Seafood and Freshwater Toxins: Pharmacology, Physiology, and Detection, *edited by Luis M. Botana*
104. Handbook of Nutrition and Diet, *Babasaheb B. Desai*
105. Nondestructive Food Evaluation: Techniques to Analyze Properties and Quality, *edited by Sundaram Gunasekaran*
106. Green Tea: Health Benefits and Applications, *Yukihiko Hara*
107. Food Processing Operations Modeling: Design and Analysis, *edited by Joseph Irudayaraj*
108. Wine Microbiology: Science and Technology, *Claudio Delfini and Joseph V. Formica*
109. Handbook of Microwave Technology for Food Applications, *edited by Ashim K. Datta and Ramaswamy C. Anantheswaran*
110. Applied Dairy Microbiology: Second Edition, Revised and Expanded, *edited by Elmer H. Marth and James L. Steele*
111. Transport Properties of Foods, *George D. Saravacos and Zacharias B. Maroulis*
112. Alternative Sweeteners: Third Edition, Revised and Expanded, *edited by Lyn O'Brien Nabors*
113. Handbook of Dietary Fiber, *edited by Susan Sungsoo Cho and Mark L. Dreher*
114. Control of Foodborne Microorganisms, *edited by Vijay K. Juneja and John N. Sofos*
115. Flavor, Fragrance, and Odor Analysis, *edited by Ray Marsili*

116. Food Additives: Second Edition, Revised and Expanded, *edited by A. Larry Branen, P. Michael Davidson, Seppo Salminen, and John H. Thorngate, III*
117. Food Lipids: Chemistry, Nutrition, and Biotechnology: Second Edition, Revised and Expanded, *edited by Casimir C. Akoh and David B. Min*
118. Food Protein Analysis: Quantitative Effects on Processing, *R. K. Owusu-Apenten*
119. Handbook of Food Toxicology, *S. S. Deshpande*
120. Food Plant Sanitation, *edited by Y. H. Hui, Bernard L. Bruinsma, J. Richard Gorham, Wai-Kit Nip, Phillip S. Tong, and Phil Ventresca*
121. Physical Chemistry of Foods, *Pieter Walstra*
122. Handbook of Food Enzymology, *edited by John R. Whitaker, Alphons G. J. Voragen, and Dominic W. S. Wong*
123. Postharvest Physiology and Pathology of Vegetables: Second Edition, Revised and Expanded, *edited by Jerry A. Bartz and Jeffrey K. Brecht*
124. Characterization of Cereals and Flours: Properties, Analysis, and Applications, *edited by Gönül Kaletunç and Kenneth J. Breslauer*
125. International Handbook of Foodborne Pathogens, *edited by Marianne D. Miliotis and Jeffrey W. Bier*
126. Food Process Design, *Zacharias B. Maroulis and George D. Saravacos*
127. Handbook of Dough Fermentations, *edited by Karel Kulp and Klaus Lorenz*
128. Extraction Optimization in Food Engineering, *edited by Constantina Tzia and George Liadakis*

Additional Volumes in Preparation

Physical Principles of Food Preservation: Second Edition, Revised and Expanded, *Marcus Karel and Daryl B. Lund*

Handbook of Vegetable Preservation and Processing, *edited by Y. H. Hui, Sue Ghazala, Dee M. Graham, K. D. Murrell, and Wai-Kit Nip*

Food Emulsions: Fourth Edition, Revised and Expanded, *edited by Stig E. Friberg, Kåre Larsson, and Johan Sjöblom*

Handbook of Frozen Foods, *edited by Y. H. Hui, Paul Cornillon, Isabel Guerrero Legarreta, Miang Lim, K. D. Murrell, and Wai-Kit Nip*

Handbook of Dough Fermentations

edited by

Karel Kulp
American Institute of Baking
Manhattan, Kansas, U.S.A.

Klaus Lorenz
Colorado State University
Fort Collins, Colorado, U.S.A.

MARCEL DEKKER, INC. NEW YORK • BASEL

Library of Congress Cataloging-in-Publication Data
A catalog record for this book is available from the Library of Congress.

ISBN: 0-8247-4264-8

This book is printed on acid-free paper.

Headquarters
Marcel Dekker, Inc., 270 Madison Avenue, New York, NY 10016, U.S.A.
tel: 212-696-9000; fax: 212-685-4540

Distribution and Customer Service
Marcel Dekker, Inc., Cimarron Road, Monticello, New York 12701, U.S.A.
tel: 800-228-1160; fax: 845-796-1772

Eastern Hemisphere Distribution
Marcel Dekker AG, Hutgasse 4, Postfach 812, CH-4001 Basel, Switzerland
tel: 41-61-260-6300; fax: 41-61-260-6333

World Wide Web
http://www.dekker.com

The publisher offers discounts on this book when ordered in bulk quantities. For more information, write to Special Sales/Professional Marketing at the headquarters address above.

Current printing (last digit):

10 9 8 7 6 5 4 3 2 1

PRINTED IN THE UNITED STATES OF AMERICA

Preface

The ancient technology of sourdough fermentation—a mixed process of alcoholic and lactic acid fermentations—has encountered a revival in the U.S. baking industry as a major method in the manufacture of specialty breads and various bakery products.

Although sourdough production is commonly practiced in retail bakeries, especially in various regions (e.g., San Francisco, California) and ethnic communities in the United States, its development and processes are less advanced than in the highly mechanized large scale baking industry which uses baker's yeast as the sole biological leavening agent. Also, information regarding scientific principles and applications of this technology is not very accessible in the United States, being scattered in trade journals or supplied by manufacturers of various commercial sourdough preparations. Further, many of the basic microbial and biochemical investigations were conducted in Europe and are mainly available in foreign languages such as French, German, or Spanish.

To overcome this information gap, we designed this book to cover the broad spectrum of this technology, focusing on principles, production of sourdough starters, and their use in the baking industry, thus providing a source of information—both practical and theoretical—for baking and food technologists as well as research microbiologists and cereal chemists. We were fortunate to gain in this effort the excellent cooperation of international scientists active in this field. Consequently, the profile of the book is truly international. Of the 11 chapters of the book, four were contributed by U.S. authors, two chapters by the joined effort of U.S.–

German authors, one from Germany, two by Spanish scientists, one from Finland, and one from France.

The reliable and rapid leavening action of baker's yeast in bread doughs by the formation of carbon dioxide leads to large-scale production of this microorganism in essentially pure form and its almost universal use as a biological leavener in baking throughout the world. This development also helped facilitate the organization of wholesale mechanized bakeries.

On the other hand, sourdough fermentation is mostly used in retail operations, and the source of the starters is confined to spontaneous fermentation in the bakeries or to the use of commercial starters. The preparation of the latter is not uniform and their performances are variable.

The main differences between the baker's yeast and sourdough fermented breads are physical and organoleptic. The physical characteristics include the higher crumb compactness and heavier crusts of sourdough breads, compared with those fermented by yeast, which exhibit high loaf volume and soft crumb with uniform crumb grain and thin crust. The organoleptic properties that are attractive to some markets are the acidic and intense flavor of sourdough fermented bakery products and the highly flavorful crust. The flavor of yeast-leavened bakery products is essentially neutral and mild, especially when they are fermented for a short time.

These differences contribute to the consumer appeal of sourdough and are positive marketing parameters since consumer preferences have shifted toward more flavorful and spiced foods. The trend seems to be attributable to increased exposure of the U.S. population to foreign and exotic foods, demographic changes, and food experiences during travel at home and abroad.

The scope of the book covers microbiological, biophysical, technological, and biochemical aspects of the fermentation processes. Chapter 1 leads us through the history of fermentation from ancient times to the present. Chapters 2 and 3 cover biological fundamentals of yeast and lactobacilli fermentations; Chapter 4 reports on interaction of yeasts and lactobacilli in dough fermentation; Chapter 5 describes U.S. yeast and sourdough fermentations, including a brief outline of baker's yeast production; Chapter 6 details the microbiology and technology of the famous San Francisco sourdough bread; Chapter 7 presents sourdough fermentation in U.S. rye bread production. The remaining four chapters introduce the French (Chapter 8), Spanish (Chapter 9), German (Chapter 10), and Finnish–East European (Chapter 11) sourdough processes and investigations.

Finally, we would like to extend our thanks to the authors of the individual chapters for their effort and sharing of knowledge. We also express our hope that this work will stimulate further interest in this field and lead to new applications worldwide.

Karel Kulp
Klaus Lorenz

Contents

Contributors

Carmen Benedito Ciencia de Alimentos, Instituto de Agroquímica y Tecnología de Alimentos, Valencia, Spain

J.-M. Bruemmer Federal Center for Cereal, Potato, and Lipid Research in Detmold & Muenster, Detmold, Germany

James J. Foy* Red Star Yeast and Products, Milwaukee, Wisconsin, U.S.A.

Kati Katina VTT Biotechnology, Espoo, Finland

Heikki Kerojoki† Vaasan & Vaasan Oy, Helsinki, Finland

Karel Kulp American Institute of Baking, Manhattan, Kansas, U.S.A.

Klaus Lorenz Department of Food Science and Human Nutrition, Colorado State University, Fort Collins, Colorado, U.S.A.

Current affiliation:
*Minn-Dak Yeast Sales & Service, Division of Sensient Technologies Corporation, Milwaukee, Wisconsin, U.S.A.
†UAB Vilnaus Duona Plius, Vilnius, Lithuania.

Daniel H. Maloney* Red Star Yeast and Products, Milwaukee, Wisconsin, U.S.A.

M. Antonia Martínez-Anaya Instituto de Agroquímica y Tecnología de Alimentos (CSIC), Valencia, Spain

Bernard Poitrenaud Lesaffre International, Marcq-en-Baroeul, France

Cristina M. Rosell Ciencia de Alimentos, Instituto de Agroquímica y Tecnología de Alimentos, Valencia, Spain

Peter Stolz Consultant, Minden, Germany

T. Frank Sugihara Consultant, Tustin, California, U.S.A.

Teija-Tuula Valjakka Vaasan & Vaasan Oy, Helsinki, Finland

Ronald L. Wirtz Medical College of Georgia, Augusta, Georgia, U.S.A.

* *Current affiliation*: Medical College of Wisconsin, Milwaukee, Wisconsin, U.S.A.

Handbook of Dough Fermentations

1
Grain, Baking, and Sourdough Bread: A Brief Historical Panorama

Ronald L. Wirtz
Medical College of Georgia, Augusta, Georgia, U.S.A.

Many archeologists today believe that humans of one type or another have been present on Earth for more than 4 million years. We know very little about our most distant ancestors, but in recent times a substantial body of knowledge has been developed about the evolution of human civilization since the last ice ages. We know that is only in the past 11,000 years that we have made the change from hunter-gatherers to farmers [1], become relatively sedentary, and begun to develop the type of agriculturally based society that evolved much later into the ancient city state of Sumer, the later society of Dynastic Egypt, and eventually into the large nation-states of the present.

For thousands of years the cereal grains and the many varieties of breads made from them have been essential elements of human nutrition. Indeed, it could be argued that most human civilizations throughout history have been based on the production of grain crops. One of the first uses of grain as food in primitive societies was doubtless in the form of parched or roasted rye, wild barley, or wild einkorn wheat, and there is considerable archeological evidence in the form of carbonized grains to show that this was common practice among hunter-gatherer groups in the Middle East as early as 12,000 BC [2].

An alternative method of preparation, that of boiling crushed or pounded grains, appears to have been developed later, and was probably first practiced by dropping small heated stones into a container holding a mixture of crushed grain and water. Both the technique and the technology have undergone considerable improvement through the ages, but the consumption of porridge is still a daily ritual for a great many people throughout the world.

Some early varieties of wheat, particularly einkorn (*Triticum monococcum*), appear to have been especially suitable for porridge. They generally do not function well in breadmaking, and also tend to have a smaller relative starch content and a higher protein level than either emmer or bread wheats. Furthermore, although it was one of the first crops of the Neolithic Period, einkorn wheat does not seem to be well adapted to a very warm climate, and is more rare in archeological sites in the extremely hot southern areas of the Fertile Crescent than either barley or emmer wheat [3].

Although einkorn does not appear to have contributed directly to the development of bread wheats, it migrated with the farmers who sowed it, spreading west into Crete and Greece, and north and west into eastern and central Europe [4]. Early bread wheat, originating from the combination of emmer (*Triticum dicoccum*) with *Aegilops squarrosa*, a wild grass common in the Caucasus, also spread along the same favorable east–west climactic zone axis [5]. Louise Levathe, citing the eminent Italian archeologist Dr. Cavelli-Sforza, notes that

> "after the introduction of agriculture in the Middle East about 10,000 years ago, farmers from there spread at the rate of one kilometer, or five-eighths of a mile, a year, eventually settling throughout Europe" [6].

Human genetic evidence also supports this somewhat controversial migration theory. Researchers from the State University of New York at Stony Brook conducted a survey of the genetic makeup of persons at 3300 sites across Europe. They found a gradual dilution of certain marker genes across an east–west gradient, supporting this outmigration of human populations across the same pathways as agriculture spread across the continent [7].

Hubert Chiron states that the production of porridge doubtless also led to the fashioning of small unleavened cakes or *galettes* for baking on hot stones or in embers, which were thus precursors of both flatbreads and leavened breads [8]. Because of a tendency to spoilage and/or fermentation, both porridge and boiled grains could only be kept for a short time and were not really suitable for travel—a definite disadvantage for nomadic hunter-gatherers [9].

Raymond Calvel, the world-famous French baker and bread researcher, gives a simple illustration to show how such "spontaneous" fermentation might have occurred:

> A present-day example of this phenomenon may be observed in summer, during stormy or warm and humid weather, in the laboratory of mills equipped with a Chopin alveograph. During the trials with this apparatus, the scraps of dough used in the course of tests are thrown into a wastecan. (This dough contains no added yeast). Whenever one forgets to empty the scrap dough from the wastecan at the end of the week, it will be full on Monday morning, even though it was only one-third full on Friday. A natural inoculation by ambient yeasts has taken place, followed by the production of carbon dioxide gas and rising of the dough [10].

This tendency, especially when the grain in question was emmer, einkorn, or one of the primitive bread wheats, is likely to have been the origin of the first sourdough. How this simple, natural event evolved into the first major food processing industry in the world is a fascinating story, and one that really cannot be separated from the development of agriculture and the beginnings of society itself.

Bernard Dupaigne, in his notable work *Le pain*, notes that on the world scale during the closing years of the 20th century, fully 53% of calories consumed were still furnished by edible grains. Rice and wheat contributed 21% and 20% respectively to this total [11], and some authorities believe that the proportion claimed by wheat is even higher [12]. Historical estimates of per capita wheat and bread consumption indicate that in the past the contribution of grains to human nutrition was much greater than is the case today [13].

For the past 5000–6000 years, much of the wheat eaten by humans has been consumed as leavened bread, and until fairly recently nearly all of this bread was produced as sourdough. Thus, in many respects, the history of sourdough bread—only one of the many faces of Mother Wheat—has been an underappreciated but important facet of the development of civilization itself.

J. Storck and W. D. Teague, in their landmark 1952 historical study of flour milling, pointed out that there has been significant popular and historical agreement on the geographic origins of the cultivation of cereal crops for a very long time. They note that in Egyptian myth, the god Osiris created agriculture near the city of Nysa, in Asia Minor, and his wife, the goddess Isis, was the first to transform barley and wheat into bread. The Greek legend of the rape of Persephone, which is also a metaphor for the origins of agriculture, states that the young goddess was carried off from the plains outside the same city of Nysa, identified by the Roman historian Pliny as Sythopolis [14].

In more recent times, this same city has been known variously as Beisan, Tell al-Husn, or Bet She'an, and is today a small Israeli settlement just a few kilometers from the western bank of the Jordan River, to the south of the Sea of Galilee. It is in this same desert region, not far from Mount Carmel, that some researchers continue to place the origins of agriculture. Archeologists have found evidence that a hunter-gatherer people called the Natufians first began to selectively harvest grain near a site called Wadi Al-Natuf, also in today's Palestinian West Bank [15]. Hundreds of small Natufian settlements have been found throughout the entire region, and evidence found at two of them in particular, Ain Mallaha (Eynan) (10,000 to 8200 BC) in the West Bank and Mureybat (8500 to 6900 BC) near Aleppo in Syria, indicates that selective grain gathering and even primitive farming were practiced there on an ongoing basis [16]. This is more than 2000 years before the farming societies of Fayum, in Egypt, or of Sialk, in Iran, where earlier researchers believed they had found some of the oldest evidence of agriculture [17].

However, it is highly likely that there were other hunter-gatherer groups even before or contemporaneous to the Natufians within this general region, for other villages have been found in Anatolia and elsewhere in Asia Minor where the inhabitants were harvesting primitive wheats, barley, and other seeds by 8000 BC [18]. Indeed, recent evidence suggests that rye was grown by semisedentary hunter-gatherers in the Near East much earlier than that. Archeological research conducted at Abu Hureyra, near Aleppo in Syria, shows that rye was being cultivated there as early as 13,000 years ago, or 11,000 BC, while work at Ohallo II in Israel indicates that wild grains formed a major part of the diet of the inhabitants of that site by 17,000 BC—even before the change to a colder, dryer climate brought on by the Younger Dryas Ice Age [19].

Given that the origins of agriculture and grain farming point to a much more ancient origin than previously believed, it would be almost impossible to state with any degree of accuracy who were the actual inventors of bread or exactly where it was first developed. Archeologists have found unleavened *galettes*, made from a variety of wild grains and baked under embers or on heated stones, in prehistoric sites as far apart from one another as the island of Crete, in Palestine, and in Iraq [20]. It is a curious fact that even the Australian Aborigines, as isolated as they were from foreign influences, have known for eons how to grind or pound wild seeds and grains to make a dough that was made into small dough cakes and baked in embers [21].

As we have noted previously, this primitive breadmaking was almost certainly preceded by roasting whole wheat and barley grains or other seeds or by boiling them, either whole or in crushed or ground form. The latter method, in addition to allowing consumption of grain in the form of porridge, also led to the development of both beer brewing and leavened bread. We know that beer was considered both a food and a beverage in Egypt, and that leavened or unleavened bread was the most basic single food item in the diet of the average Egyptian. For many years it was popularly believed—and often repeated—that leavened bread developed in Egypt and spread gradually to other parts of the Middle East and then to Europe [22]. Archeological evidence discovered in 1983 in Switzerland, however, shows that the production of leavened bread had already reached a high state of development in that area by at least 5800 years ago, i.e., more than 3800 years BC [23]. This is perhaps another piece of evidence to support the Cavelli-Sforza outmigration theory, which in much simplified form states:

> This record indicates agriculture being practiced in eastern and central Turkey around 10,500 years ago, in western Turkey between 7,500 and 8,000 years ago, in southern Europe between 7,000 and 7,500 years ago, Central Europe 6,000 to 6,500 years ago, France and north Germany 5,500 to 6,000 years ago, Sweden and Russia 5,000 to 5,500 years ago, and in the British Isles and Finland between 4,500 and 5,000 years ago [24].

This does not mean that the honor of having developed breadmaking to the technology stage does not belong to the Middle East, for it was there that baking gradually advanced to the level of being able to provide bread on a very large scale, while in Europe and elsewhere the consumption of grain as porridge remained common well into the 13th century [25].

In Babylon there were already primitive ovens around 4000 BC [26]. In general, bread eaten by the common people of Babylon or of pre-Dynastic Egypt were unleavened, made at the household level from barley flour, and hardly larger than *pistolets* or small dinner rolls today [27]. By the early Dynastic Period, the skill and ingenuity of Egyptian bakers had considerably expanded the bread choices available. H. E. Jacob notes that Egyptian bakers made more than 50 different types of cakes, unleavened breads, and breads leavened with beer foam or with sourdough [28]. Many of these were made from barley, but others were made from fine white wheat flour, and some were even fried in oil, much like modern *beignets* or doughnuts. However, these finer products were destined to the privileged classes, while the common people had to be satisfied with common barley bread and hearth cakes.

The technology of the large-scale manufacture of leavened bread is still attributed to the Egyptians, and it is important to recognize the significance of the industrial scale production of leavened bread to the development of Egyptian society. The great bakeries of the pharaohs were remarkable for a high degree of division of labor, and the use of large batteries of ovens made it possible to reach very high production levels. The scale of production is still surprising to us today. A large bakery, capable of producing enough bread to feed more than 15,000 workers per day, was discovered in 1991 within sight of the Great Pyramids [29].

It is apparent that the development of agriculture in the Nile River valley, especially the cultivation of wheat and other grains, combined with progress in the technology of bread production, made it possible to produce large agricultural surpluses and to convert them to nutritious food on an efficient industrial scale. This also made possible the creation of a stable, centralized government, and led indirectly to the other technological, cultural, and intellectual accomplishments of Egyptian society [30].

The diet of the ancient Greeks was originally based on meat and milk products from their flocks, supplemented by unleavened barley bread and grain porridge. Because of the climate and the poor, rocky soil of their homeland, they were obliged at an early date to import large quantities of barley and wheat from Egypt, Sicily, and even the Crimea [31]. It was also from the Egyptians that the Greeks learned to make leavened bread, principally sourdoughs, although some types were leavened with beer yeasts, and the Greeks also prepared a dried leaven from wine lees. Boiled porridge and unleavened barley bread remained the principal foods of the lower classes and slaves, but a variety of breads and pastry products were to be found on the tables of the rich, the nobility, and members of the wealthy artisan class.

Skilled Greek bakers are known to have made at least 72 different varieties of bread, including types of fine white bread, cheese breads, barley and rye breads, and even variety breads flavored with honey or anise, which might also be decorated with poppy or sesame seeds [32]. They also made biscuits [33], cheese and honey cakes, and a variety of fruit-filled pastries. There were many different types of *pains de fantaisie*, including braided loaves, breads in animal forms, ring-shaped loaves, and elongated "baton" loaves [34]. Again, the majority of these were produced as sourdoughs, especially in the case of more common breads.

Just as had occurred in the arts, philosophy, and the sciences, the Greeks had made great strides in the technology of baking. Hubert Chiron notes that Chrysippus of Diane wrote the first technical book on baking in history, dating from around 1900 BC [35]. Even though this document has disappeared since that time, it is evident that Greek bakers played an important role in the transformation of the national diet, on the development of commodity trade both within and without their country, and thus in the transformation of Greek society from a loose confederation of sheepherders, fishermen, and small farmers into a resolutely urbanized and sophisticated society. The influence of Greek bakers was also evident in the advances they made to baking equipment. They improved on the conical Egyptian oven by developing a sectioned dome, thus making it more efficient and capable of a higher level of output. The noted baking science author and editor E. J. Pyler has pointed out that the Greek oven served as the model for those used throughout the civilized world until the industrial revolution [36]. The Greek rotating saddle mill or *quern* also had a great deal of influence on milling technology until the development of the modern rolling mill. It is a curious fact that these anonymous bakery innovators, who did so much to transform the food and agricultural systems of an entire civilization, were almost all slaves.

Greek slave bakers were also very important to the great changes that occurred in the food systems of the people of Rome, since by changing their basic food they also transformed the structure of their society. Bernard Dupaigne notes that

> L'histoire du pain à Rome est de toute première importance pour qui veut comprendre le destin de l'Empire romain, son ascension glorieuse, et sa chute [37].

The primary food of the early Romans was boiled grain porridge. However, from the 3rd century BC on, under the influence of the refinements of Greek philosophy and civilization, rich Romans began to employ Greek slave bakers in their homes. After the defeat of Perseus, king of Macedonia, in 168 BC, many Greek bakers were taken to Rome as slaves [38]. Bread quickly became a staple food, and from 173 BC on there were professional bakers [39], some of whom were liberated slaves who had become rich [40], while others were respected magistrates [41].

By the time of the Emperor Augustus there were 329 public bakeries in Rome, which at that time had more than a million inhabitants [42]. At Pompeii there were more than 20 bakeries for 20,000 persons [43]. The production levels of the largest Roman bakeries were impressive. The bakery of the Latini brothers, for example, milled more than a thousand bushels of wheat and produced 150,000 loaves of bread per day [44]. Under the Roman Empire, the bakery profession was respected and even encouraged by the emperors, as the baker had become indispensable to a society that principally nourished itself with bread. The members of the *corpus pistorum*, the bakers union, obtained numerous special rights and privileges as a sign of their importance to the well-being of the nation [45], but little by little the independent artisan baker was transformed into a public official, and the profession of baker became hereditary and obligatory. The Empire needed them to ensure the food supply for a citizenry who lived in a permanent condition of state-supported indolence, and baking was the most efficient way to produce food on such a large scale.

By the time of Julius Caesar, around 46 BC, half of the citizens of the city of Rome—more than 300,000 of a population of 600,000 free citizens—received free wheat from the public granaries [46]. Three hundred years later, during the reign of the Emperor Aurelian [47], the state had begun to distribute bread three times per day in the place of wheat. In effect, the establishments of the great bakers of Rome had been nationalized; the bakers and their apprentices had become employees of the state, and sourdough breads nourished the majority of Roman citizens [48].

The Barbarians who brought an end to the Roman Empire in 476 had practiced subsistence agriculture for thousands of years, but bread was not one of their primary foods [49]. Boiled grains, acorns and other nuts, a few types of legumes, and other vegetables became the bases of the average diet [50]. Under the periodic assaults of successive waves of Barbarian migration, the agricultural and food production practices of the Gallo-Roman civilization of Gaul and Provence entered into a decline that lasted until the 10th century AD. Industrial scale baking as practiced by the Egyptians and the Romans disappeared completely. The techniques and technology of agriculture, milling, and baking survived primarily in the larger abbeys and monasteries [51]. It was not until the 12th century that the profession of *fournier*—the first bread baking specialist—began to appear in the larger towns. Another bread-related speciality—that of *tamelier* or sifting and mixing specialist—also began to appear [52], and these two groups joined forces in France to establish the profession of "baker" by purchasing a corporation license or charter from the King in 1260 [53].

The "new" baker's profession was very closely regulated by laws and rules on the quality, weight, and price of bread. In 1266, Henry III of England proclaimed the Assizes of Bread, thus establishing the price of bread in relation to the price of grain, and limiting the baker's profit to no more than 13% [54]. By 1372, the French baker had the right—at least in principle—to make three different types of bread, one corresponding to each of the three social classes [55]. In his *Glossaire de la basse latinité*, Du Cange noted that 20 different varieties of bread were served at the King's table [56], and it is surmised that all of these were different types of sourdough. Only members of the nobility could afford the highest quality white bread, a situation that had hardly changed since the days of the ancient Egyptians [57]. The bourgeois ate gray or brown sourdough breads, often made from *maslin*, a blend of wheat and rye flour [58]. The common people had only ball-shaped loaves of black bread, made from a mixture of rye, barley, and wheat flour [59]. It was by association with this most common and widely consumed type of bread, the black ball-shaped sourdough loaf, or *boule*, that the French baker received the name of *boulanger* [60].

Throughout the later Middle Ages, aided by feudal ordinances that required milling of grain in the lord's mill and baking in the *four banal* or lord's oven, bread was once again reestablished as the staple food of the village-dwelling peasant. The social status of the baker also began to rise as his importance to maintenance of public order was again recognized, much as it had been in Rome [61].

Bread eventually was so important that the professions of baker and miller became very hazardous during the frequent periods of famine, since starving crowds often sacked bakeries and mills and assaulted or killed mill operators and bakers. Starvation was a constant danger to the common people throughout Europe. Since the fall of Rome the continent had often been devastated by war, especially the Hundred Years War, by plague, by extremely hard winters during the Little Ice Age from about 1350 to 1450, by disastrous harvests, and by religious upheaval. When wheat, barley, or rye was not available, bread was made from acorn flour, wild grass seeds, pine bark mixed with flour, earth mixed with flour, or animal blood mixed with barley flour. Centuries later the great Parmentier, Minister of Agriculture under Napoleon, remarked that he did not understand the "mad desire to turn everything into bread" [62]. When even these strange concoctions were unavailable, it was not uncommon for men to turn to cannibalism, a seldom discussed situation that existed for several hundreds of years in eastern Europe [63].

Sourdough bread, the end product of the most widespread technology of agriculturally based civilization, had become essential to the survival of all other elements of society, for without bread men reverted to the state of beasts. The baker, just as much as the monk, the humanist scholar, or the architect of the great cathedrals, contributed greatly to the social progress made during the difficult period of the Middle Ages.

Following the reemergence of towns during the later Middle Ages, the technology of breadmaking that had been conserved by monks since the end of the Roman Empire began to make new progress throughout Europe. This was due in part to the formal system of apprenticeship that was practiced by the professional corporations or guilds. Improvements in bread quality were stimulated by the exchange of knowledge that journeyman bakers acquired during the several years that they were required to spend with different masters, sometimes in different regions or countries. Improvements were also made in mill and oven design for the same reasons:

> The guild rule that forced the journeyman to ultimately return to his home town relieved the employing master of any anxiety over eventual competition, so that he usually was more than willing to impart his knowledge to his temporary employees. Thus the art of baking, which

> attained great heights in such cities as Vienna and Paris, was constantly nurtured by an interchange of experiences on a truly continental scale [64].

The influence of Catherine and of Marie de Medici on the evolution of literary and artistic taste in the court of the King of France is well known. The contribution of their cooks and bakers is not as well recognized, but it was considerable. The bakers brought back to France the practice of using beer yeast as a leavening agent in bread production, thus providing an alternative to the use of the *levain* or sourdough method that had been in use almost exclusively for more than 1000 years.

The use of beer yeast in breadmaking had been practiced by the Gallo-Roman bakers at the time of Julius Caesar but had been long since forgotten. It is thought that this ancient practice had been imported from the Middle East by the Phonecians and transplanted to Marseille around 600 BC, but along with many other technological practices this had disappeared following the fall of the Roman Empire [65]. Thus, the nobility and wealthy bourgeois again had access to high-quality, light white bread. The common people continued to consume its customary sourdough (*pain au levain*), most often bran or barley bread, but the Italian bakers had implanted in France a taste for white bread that would have unexpected and even "revolutionary" consequences two centuries later.

The 17th century was incredibly rich in intellectual and scientific progress, but much less rich in grain suitable for bread. There were terrible regional famines and, as during the Middle Ages, it was necessary to make bread from ingredients such as vegetables, dried fish, acorns, roots, and even ground clay tiles [66]. However, all of these problems were not caused by unfavorable climactic conditions, as in the past. Excessive taxation levied on the peasants, which reached as high as 80% of the value of production under Louis XIV and his successors, were a primary cause of this privation:

> The French peasantry at the height of Louis Quatorze's glory were seen in the neighborhood of Blois eating nettles and carrion. Women and children were found lying dead by the roadsides, their mouths stuffed with inedible weeds. Madmen crouched over graves in cemeteries, sucking and gnawing at bones. ... In 1698 various governors reported that France was beginning to die of starvation; famine swept over it unchecked [67].

The peasant, deprived of bread and even of the tools necessary to grow grain and other foods, often turned to theft. Bands of starving peasants pillaged the grain stocks of the chateaux of the rich nobles, and captured carts and barges loaded with grain [68]. In France, more than 100,000 people were condemned to life imprisonment on the royal galleys [69]. As many as 50,000 persons were arrested in France in a single day [70]. The social disintegration that led to the French Revolution had already begun, precipitated as much by the indifference and incomprehension of the Court as by hunger.

It was only in the 18th century that scientists and technicians finally began to make the first coordinated efforts to combat hunger through the application of knowledge and technology. In 1765 Malouin published *L'art du meunier, du boulanger, et du vermicellier*, the first technical work on milling and breadmaking since the lost work of Chrysippus de Diane. Malouin's work is important in its own right, but also as part of the movement that culminated in Diderot's great encyclopedia, the first comprehensive effort to record the totality of man's scientific and technical knowledge. A new three-stage milling process called *mouture économique* was developed in 1760 in Paris by Malisset, resulting in the production of 15% more flour per weight of wheat than had been possible previously. Even though the flour produced was also demonstrably better for breadmaking, the system was not adopted for another 25 years, due the primitive state of most French mills and the immense investment that was necessary to initiate the new milling technology [71]. Unfortunately, milling was not the only food production technology that was behind the times in France:

> The crisis in French agriculture continued throughout the 18th century, through the end of the reign of Louis 14th and during that of Louis 16th. Grain yields were still very unsatisfactory, at the rate of six for one, compared to eleven or twelve grains harvested for each one sown in the Flanders area of Belgium [72].

The famous scientist Parmentier has spent years in efforts to find a remedy for the terrible successive famines in France, and believed that he had found an answer in the cultivation of the potato and the generous dedication of French bakers. During the Seven Years War, Parmentier had been a prisoner in Germany and had noted the importance of the potato to the success of the Prussian army. On his return to France, he published *Le parfait boulanger: traité complet sur la fabrication et le commerce du pain* [73], in which he advocated the use of potato flour as an effective weapon against famine. Since Parmentier also believed that bread was "the first and most indispensible of foods" and felt that the social rank of the baker should correspond to the critically important nature of the food they produced, he also established a free *Académie de Boulangerie*, which was supported by the State and dedicated to the scientific study of bread made from mixed or composite flours as well as from wheat. Unfortunately, his efforts were frustrated by the fearful and obstinate attitude of the people; some were afraid that that potatoes caused illness, while others refused to eat food that was normally fed to animals. His proposals were even attacked by the ruling classes, who severely criticized the taste of bread made from a mixture of wheat flour and potato flour. Parmentier responded in his work, *Traité sur la culture et les usages de pomme de terre* [74], but his work was in vain. It was too late to repair the harm caused by a century and a half of poor agricultural practices and unbridled economic inflation [75]. In spite of the best efforts of reformers, philosophers, scientists, and even millers and bakers, the French revolution was ready to explode.

It was an additional tragedy that the political upheavals occurred had only a minimal effect on a series of famines that succeeded one another almost without a pause from 1789 to 1796. Even the adoption of bread rationing imposed by the Convention had little effect [76]. Political idealism may have even made the situation much worse, since it led to the adoption in 1793 of the production of "*pain d'Égalite*," white bread that for a time was virtually all that could be legally and produced and sold—formerly such white bread had been available only to the upper classes. The Revolution dictated that it should be equally available—or unavailable—to all citizens:

> France saw no bread until peace came. The Revolution had not been able to produce it, and the war made it impossible to distribute it. It was not until the Directory, from 1796 on, that the soldiers were furloughed; they returned to the fields which now no longer belonged to landowners but to themselves and their families, and they began to till these fields [77].

In the long run this great social upheaval accomplished remarkable changes in agriculture in France, while at the same time serving as an example for other revolutions around the world. Whether in the name of equality or for less ideological reasons, the former soldiers who had become farmers planted little grain other than wheat. As a result, white wheat sourdough bread almost totally replaced other types of bread in France, except in parts of the Massif Central and the Limousin where the climate and soil were less suitable for wheat [78]. Continuing progress in the science of milling in England and the United States also contributed to the dominance of white wheat flour over that made from other types of grain [79]. Parmentier's continuing efforts to promote the cultivation of the potato were more effective due to his status as minister of agriculture under Napoleon. This resulted in a considerable decline in the consumption of other grains in France, as well as a more gradual downward trend in the total consumption of wheat flour and wheat sourdough breads [80].

As the various state economies began to recover from the ravages of revolution and war, a number of other changes occurred in the food habits of the average western European. French bakers perfected the production of *sur levain* (sourdough) white breads due to the availability of higher quality wheat flour. They also began to adopt the use of beer yeast in bread production on a larger scale, especially in winter. Austrian bakers who emigrated to Paris around 1840 initiated the production of Vienna breads and other luxury products. A Polish nobleman, the Baron Zang, introduced the use of the *poolish*, a multistage fermentation method based on the use of prepared yeast that is still practiced by specialists today [81]. Under the impact of changing economies and evolving tastes, the traditional rounded form of the bread loaf changed to an elongated form with a longitudinal split [82].

From about 1850 on, improvements in transportation made it possible to import wheat in quantity by ship from North America or Russia as needed, and to deliver it economically by rail. By 1900 there were more than 80,000 artisan bakeries in France, 90,000 in Germany, and 2000 in Belgium. This was truly the "golden age" of artisan bakers, but the artisan period was limited by advances in technology and the mechanization of bread production. Throughout Europe, the early 20th century saw a gradual increase in the adoption of mechanically assisted production methods, even in the traditionally oriented baking industry, and artisan production of bakery foods experienced a gradual decline throughout western Europe until relatively recently [83].

Competition from larger scale industrial producers, including on-site bakeries in hypermarkets and supermarkets that utilize frozen dough or partially baked, fully proofed and formed dough, continues to pose a serious problem for traditional artisan bread bakers. The situation is rendered more complex by significant improvements in frozen dough technology, combined with a general lack of consumer knowledge and sophistication regarding quality attributes of breads. In France and Italy it has even become necessary to protect artisan producers and products by defining precisely what is and what is not a traditional sourdough or artisan bread.

Except for a very few minimally fermented flatbreads based on such grains as corn or quinoa, bread production in North America had a much more recent start, and evolved much more rapidly from a traditional to a highly mechanized production basis than in Europe. The indigenous inhabitants of the Americas had developed food systems that were rich and productive, but quite foreign to the earliest European settlers. Efforts to introduce wheat culture were made as early as 1529 by Spanish colonists in Mexico, and the English first planted wheat at their colony in Roanoke in 1585 [84]. A mere half-century later, there was already a grain surplus in Connecticut, and a single producer was able to export more than 1500 bushels of wheat in a year. By 1631 the city of New York had become an important commercial milling center, and by 1678 wheat exports had reached 60,000 bushels per year. Whereas in France the price of bread had risen by 300% in 1765 alone [85], Philadelphia grain merchants exported 365,552 bushels of wheat, 148,887 barrels of wheat flour, and 34,736 barrels of hard ships' biscuit that same year [86]. North America was truly the "land of plenty," although there were some difficult periods during the War of Independence:

> There was always sufficient food in America to feed the Continental troops bountifully; transportation and mismanagement, most of which were avoidable, kept the army nearly always in want. The quantity of the bread ration was cut down many times to eke out the supply during periods of scarcity. Several times during the year 1779, and not always during the winter months, the Northern Department troops were on the verge of mutiny for lack of bread [87].

In comparison to the European experience, periods of real famine have never really existed in North America. Personal liberty and self-determination in combination with rich agricultural resources have made it possible to produce significant surplus of food and other agricultural products on a frequent basis. The early United States had an abundance of natural resources but a shortage of workers, with the result that Americans were obliged to multiply the efforts of their efforts by adopting technology at a relatively early date. In comparison to the primitive flour mills in general use in France during that same period, the mill constructed by Oliver Evans in 1785 was capable of operating for significant periods of time without any human intervention [88].

The latter half of the 19th century saw an acceleration of the transformation of bread baking in the United States from a large number of traditionally oriented small shops to a large-scale mechanized industry, a trend that began in the biscuit and cracker baking industry nearly a century before [89]. This change was driven by advances in the technology of mechanical engineering, which allowed the production of mixing machines, large-capacity reel and traveling ovens, and other advances in the technology of baking; by the increasing concentration of the U.S. population in urban centers, which made local distribution of short shelf-life bakery products practical; and by the development of economical means of both intercity (railroad) and intracity (electric and gasoline-powered) delivery vehicles. Thus, by the very beginning of the 20th century, conditions were set for a major shift in the means of production and distribution in North America, and the manufacture of white yeast–leavened breads dominated the bread market. Virtually the only exceptions were specialty ethnic and rye products.

These changes in the machine technology of American baking were paralleled by developments in the conditions of work and significant shifts in product type, quality, and sanitation. Before the machine bakery era, production areas of bakeries in both Europe and the United States were normally located in poorly ventilated basements, with inadequate sanitary controls and poor or nonexistent attention to the quality and purity of ingredients. Proper sanitation was given minimal attention, because bakery personnel often lived in the same areas in which bread was produced. Bakery products made under such primitive conditions were often considered to be substandard, and home-baked products were thought to be more healthful and more nourishing.

The adoption of technological advances in North American baking was greatly stimulated by the First World War of 1914–1918. Government-sponsored programs had enabled the large-scale baking industry to solve difficult ingredient shortage problems by cooperative effort and the crash application of technology to production of bakery products, primarily bread. These changes were not adaptable to smaller scale production, with the result that many traditionally oriented small shops went bankrupt during the war and during the economically difficult 1920s and 1930s [90]. There was a consequent reduction in the different types of products available to the consumer, and this was furthered by a "patriotic" wartime reaction against German-type products such as sourdough rye breads—a situation that occurred again during the Second World War.

Due to the important involvement of women in industry, the production of baked goods in bakery plants grew from approximately 15% of total consumption in 1913 to nearly 35% by the war's end, and many families adopted the practice of purchasing bread from bakeries or grocery stores, a habit that they did not quickly abandon after the end of the war and the return to normalcy [91]. In a sense, commercially produced yeast-leavened bread became the original "convenience food" in the United States during and immediately following the First World War.

The impact of food shortages and restrictions on the U.S. baking industry was less during the Second World War than during the First World War, due in part to enlistment of industry under government orders and the commitment of resources on a total war footing [92]. In addition to reduced stocks of wheat flour, there were reductions in fats, oils, and shortenings; reduction or elimination of cellophane and special wrappings; and extension of the working life of baking machinery far beyond normal. In fact, there was essentially no replacement of baking machinery for the duration of the war.

In the face of these shortages and restrictions, the baking industry was obligated to actually increase the nutritional value of bread and bakery products and to achieve greater economies of production than ever before. Due to scarcities in other foods, consumption of wheat products in the American diet increased from 25% in 1939 to 40% in 1944. The number of bread and cake bakers in the United States increased by 27% from 1943 to 1944 alone, with a gain of 65% in production volume during that same period. In that single year white bread production increased by 11% and rye bread by 9%. A survey conducted in 1944 indicated that large commercial bakers supplied 85% of all bread and 35% of the cake consumed in that year, although profits were lower than in 1943 because of increased ingredient costs [93]. This growth in quantity was achieved at the expense of diversity, and many types of specialty traditional products disappeared almost entirely. Part of this change was due to the same type of patriotic reaction against ethnic products, many of which were traditional sourdoughs. However, another factor that is little known today was a widespread campaign sponsored by the U.S. government to promote the consumption of the factory-produced "American loaf," essentially an unsliced composite-flour yeast-raised "white" bread that could be made on automated equipment with unskilled labor and with tremendous economies of scale [94]. Postwar years saw the reintroduction of slicing and packaging and the return to higher quality ingredients following the economies of the Marshall Plan years, but the bread consumption habits that Americans had acquired during the war years persisted as the "baby boom" generation grew into adulthood.

This situation began to change slowly in the 1980s, however, and today there is no doubt that sourdough products and other "traditional" European-type products have experienced a renaissance in popularity over the last decade. It is now possible to find quality sourdough breads of many types in large and mid-sized cities throughout North America, although in some regions—notably south of the Mason-Dixon line—the popularity of this and other traditional breads still lags sadly behind that found in other areas of the continent. There are a number of reasons for this growth in popularity, including wider exposure of North Americans to the cuisines of other cultures through travel, a new wave of immigration from bread-oriented central and eastern Europe, continued long-term growth in disposable income among the middle and upper classes (at least until the instability in the stock market that began to occur early in 1999) and a growing concern and interest in health and nutrition matters among aging baby boomers. This latter trend has resulted in tremendous growth in consumer adoption of many different types of alternative and "traditional" systems of nutrition and health maintenance, affecting not only our food production and distribution systems but also the nation's medical and pharmaceutical industries.

Some self-styled experts within the alternative nutrition movement decry the consumption of wheat and other grain products, continuing the unfounded attacks on bread and carbohydrate consumption that were so frequent in the early part of the 20th century. However, the majority of certified nutritionists today recommend the adoption of dietary practices that are strikingly similar to "traditional" food consumption patterns. These recommendations approach dietary practices that have been common in less industrialized societies for centuries.

Although the food consumption recommendations in the *Mediterranean Diet Pyramid* [95], the *Food Guide Pyramid* developed by the U.S. Department of Agriculture and the Department of Health and Human Services, the *Australian Guide to Healthy Eating*, and *Canada's Food Guide to Healthy Eating* may differ in other details, they all agree substantially on the important role that consumption of complex carbohydrates should play in human nutrition. The emphasis of all of these programs is a return to increased consumption of bread and other grain-based products as a means to better health. They fail to take into consideration one other very important factor, i.e., products made by traditional sourdough and other extended fermentation processes are normally superior in taste, shelf life, and eating characteristics when compared with the great majority of mass-produced conventional and short-time yeast-leavened breads [96].

REFERENCES AND NOTES

1. Price, T.G. & Gebauer, A.B. *Last Hunters, First Farmers*. Santa Fe, NM: School of American Research Press, 1995, p. 3.
2. Price & Gebauer, p. 55.
3. Zohary, D. & Hopf, M. "Domestication of plants in the Old World." 2nd ed. Oxford: Clarendon Press, 1994, p. 36.
4. Diamond, J. "Location, Location, Location: the First Farmers." *Science*, Nov. 14, 1997, p.1244.
5. Sokolov, R. "The Unknown Bioengineers." *Natural History*, Vol. 102, No. 10 (Oct. 1993), pp. 104–107. See also: Allard, R.W. "History of Plant Population Genetics." *Annual Review of Genet*, v. 33 (1999), p. 2.
6. Holden, K. "Geneticists Trace Migration of Mideast agriculturalists to Europe." *Washington Report on Middle East Affairs*, Sept. Oct. 1993, p. 82. (Cites Louise Levathe's article printed in the *New York Times* July 27, 1993).
7. Sokal, R., Oden, N. & Wilson, C. *Nature*, May 1991.
8. Chiron is a respected expert on baking technology with the French Institut National de Récherché Agronomique (INRA), and one of the most knowledgeable authorities on the history of bread and baking in the world today.

9. Chiron, H. & Godon, B. "Historique de la panification" in *La panification française*. Paris: Lavoisier Tec-Doc, 1994, p. 4.
10. Calvel, R. *The Taste of Bread*. Gaithersburg, MD: Aspen Publishers, c 2001, p. 21. It should also be remembered that wheat normally has yeast spores on the outer surface of the kernel and that some of this yeast inevitably finds its way into flour.
11. Dupaigne, B. *Le pain*. Paris: La Courtille, 1979, pp. 15–16.
12. Zohary, D. & Hopf, M. "Domestication of plants in the Old World" 2nd ed. Oxford: Clarendon Press, 1994.
13. Calvel, p. 188.
14. Storck, J. & Teague, W.D., *Flour for Man's Bread*. Minneapolis, MN: Univ. of Minnesota Press, 1952, pp. 27–29.
15. Holden, Kurt. "Geneticists trace migration of mideast agriculturists to Europe." *Washington Report on Middle East Affairs*, Sept.–Oct. 1993, p. 82.
16. Moore, A.M.T. *The Neolithic of the Levant*. Oxford, UK: Oxford Univ. Press, 1978, pp. 99–106.
17. Storck & Teague, pp. 27–29.
18. Ziehr, W. *Bread through the ages*. Tielt, Belgium: Lannoo, 1987, p. 11.
19. Pringle, H. "The Slow Birth of Agriculture." *Science*, Vol. 282, No. 5393, Nov. 20, 1998, p. 1446.
20. Luraschi, A. *Il pane e la sua historia*. Torino: Edizioni de "L'Arte Bianca, 1953, pp. 9–13.
21. Dupaigne, p. 16.
22. Note: as an example, see: "On a Roll: the Sweet Success of San Francisco's Sourdough." *Image*, Jan. 18, 1987, p. 27.
23. Waehren, M. "Le monde du pain il y a environ 5500–5800 ans," *Le boulanger-pâtissier*. Paris: La Press Corporative Française, LII, 1 (January 1983), pp. 52–58.
24. Holden, p. 83.
25. Chiron, H. *Le pain français a travers les ages*. Paris: Union des Fabricants Francais d'Equipments pour la Boulangerie et la Patisserie (1998), p. 3.
26. Quarooni, Ponte, Posner. "Flat breads of the world." *Cereal Foods World*, Dec. 1992, p. 863.
27. Chiron, H. "L'histoire du pain de l'antiquité à nos jours." *INRA Mensuel*, No. 2, 1990.
28. Jacob, H.E. *Six Thousand Years of Bread*. New York: Lyons and Burford, 1997, p. 27. (Originally published by Doubleday, 1944).
29. *Modern Baking*, Feb. 1992, p. 21. See also: Wood, E. *National Geographic*, Jan. 1995.
30. Storck & Teague, pp. 54–55.
31. Jacob, p. 48; Ziehr, p. 84.
32. Dupaigne, p. 30.
33. *Ibid.*, p. 30
34. Ziehr, p. 84.
35. Chiron, H. "Historique de la panification française." *INRA Mensuel*, No. 2, 1990, p. 4.

36. Pyler, E.J. *Our Daily Bread.*
37. Dupaigne, p. 33. An appropriate English translation might read: "The history of bread in Rome is of the greatest importance for whoever seeks to understand the destiny of the Roman Empire, its glorious ascension, and its eventual fall."
38. Chiron, p. 4. Dupaigne, p. 32.
39. Ziehr, p. 86.
40. Storck, p. 85. Jacob, p. 78.
41. Pyler, p. 19.
42. Dupaigne, p. 32. Pyler, p. 17.
43. Storck, p. 84. N.b.: Dupaigne, p. 33, states 40 establishments.
44. Pyler, p. 18.
45. Chiron, p. 5. Jacob, p. 79.
46. Pyler, p. 19. See also Ziehr, p. 87, Storck, p. 89.
47. Aurelian reigned from 270 to 275 AD.
48. Dupaigne, p. 35.
49. It is certainly not the situation with the descendants of the Frankish and Germanic tribes today. The French are renowned for their skill in the production of white hearth breads, while the expertise of the Germans in the realm of rye and mixed grain breads is recognized worldwide, and there are more than 200 varieties of bread produced in Germany today. Among the countries that make up the European Community today, however, they rank in sixth and fourth places in terms of bread consumption. See: Assire, J. Le livre du pain. Paris: Flammarion, 1996, p. 171, and Calvel, p. 188.
50. Dupaigne, p. 5.
51. Pyler, p. 21.
52. Chiron, p. 6.
53. Dupaigne, p. 36; see also Ziehr, p. 89.
54. Jacob, p. 137; Dupaigne, p. 36.
55. Dupaigne, p. 36.
56. Jacob, p. 144.
57. Chiron, p. 7.
58. Dupain, p. 36; Bailey, pp. 41–42. This is called *métil* in French.
59. Dupaigne, p. 36; Jacob, p. 144.
60. Chiron, p. 7.
61. Jacob, p. 136; Pyler, p. 22; Chiron, p. 7; Bailey, p. 40.
62. Chiron, p. 9.
63. Jacob, pp. 147–148.
64. Pyler, p. 22. See also: Jacob, p. 137.
65. Chiron, p. 10.
66. Chiron, p. 10.
67. Jacob, p. 239.
68. Bailey, p. 47.
69. Dupaigne, p. 37.
70. Jacob, p. 239.
71. Jacob, p. 242.

72. Dupaigne, p. 37. The original reads: "La crise de l'agriculture française se poursuit au XVIIIe century, à la fin du règne de Louis XIV et durant celue de Louis XVI. Les rendements sont encore très médiocres, six pour un, contre onze ou douze grains récoltés pour en semé dans la Flandre belge."
73. Title translation: *The perfect baker: a complete treatise on the production and sale of bread.*
74. Literally translated as: *A treatise on the cultivation and the uses of the potato.*
75. Dupaigne, p. 37.
76. Jacob, p. 252; Chiron, p. 11; Dupaigne, p. 41.
77. *Ibid.*, p. 254.
78. Dupaigne, p. 63.
79. Storck, p. 174.
80. Dupaigne, p. 41; Jacob, p. 256.
81. Calvel, p. 116.
82. Chiron, p. 11.
83. Note: Assire, p. 171, states that there are currently about 200,000 artisan and industrial bakeries combined in the entire European Community.
84. Storck, p. 146.
85. Dupaigne, p. 37.
86. Storck, p. 151.
87. Fitzpatrick, J.C. "Christopher Ludwick: the Baker of Washington's Army," reprinted from *The Spirit of the Revolution* (New York: Houghton, Mifflin, 1924), p. 11.
88. Storck, p. 160.
89. "Cookie and Cracker Industry Has Long, Rich Past," *Snack Food Blue Book* (New York: Harcourt Brace Jovanovich, 1976), p. 16.
90. William Bradley, "Fifty Years of Progress in Conventional Baking," *Cereal Science Today* 10(6): June 1965, p. 263.
91. *Ibid.*, p. 263.
92. Walter Deitz, *AIB Commencement Address* 21 Dec. 1946.
93. *Bakers' Weekly* 9 Feb. 1942: 31.
94. This program was widely publicized by means of popular magazines and newspapers of the period, as well as by official U.S. government posters. The author has personally examined these historical materials, which are retained in the files of the Harold Freund Museum at the American Institute of Baking.
95. Mediterranean Diet Pyramid, The Oldways Preservation and Exchange Trust and Harvard School of Public Health, 1994.
96. For a discussion of the effects on taste of the various breadmaking methods, see pages 40–48 of Raymond Calvel's work *The Taste of Bread.*

BIBLIOGRAPHY

Allard, R. W. History of plant population genetics. Annu Rev Genet 33, 1999.
Assire, J. Le livre du pain. Flammarion, Paris 1996.

Bakers' Weekly 9 Feb. 1942.
Bradley, W. Fifty years of progress in conventional baking. Cereal Sci Today 10(6) June 1965.
Calvel, R. The taste of bread. Aspen Publishers, Gaithersburg, MD 2001.
Chiron, H. L'histoire du pain de l'antiquité à nos jours. INRA Mensuel, No. 2, 1990.
Chiron, H. Le pain français a travers les ages. Union des Fabricants Francais d'Equipments pour la Boulangerie et la Patisserie, Paris, 1998.
Chiron, H. and Godon, B. Historique de la panification. In La panification française. Lavoisier Tec-Doc, Paris, 1994, p. 4.
Cookie and cracker industry has long, rich past. Snack Food Blue Book. Harcourt Brace Jovanovich, New York, 1976.
Deitz, W. AIB Commencement Address, 21 Dec. 1946.
Diamond, J. Location, location, location: the first farmers. Science 278(5391), Nov. 14, 1997.
Dupaigne, B. Le pain. La Courtille, Paris, 1979.
Fitzpatrick, J. C. Christopher Ludwick: the baker of Washington's army. Reprinted from The Spirit of the Revolution. Houghton, Mifflin, New York, 1924.
Holden, Kurt. Geneticists trace migration of mideast agriculturists to Europe. Washington Report on Middle East Affairs, Sept.–Oct. 1993.
Jacob, H. E. Six Thousand Years of Bread. Lyons and Burford, New York, 1994. (Published as a reprint by Lyons and Burford 1997.)
Luraschi, A. Il pane e la sua historia. Edizioni de L'Arte Bianca, Torino, 1953.
Mediterranean Diet Pyramid, The Oldways Preservation and Exchange Trust, and Harvard School of Public Health, 1994.
Modern Baking, Feb. 1992.
Moore, A. M. T. The Neolithic of the Levant. Oxford University Press, Oxford, 1978.
On a roll: The sweet success of San Francisco's sourdough. Image, Jan. 18, 1987.
Price, T. G. and Gebauer, A. B. Last Hunters, First Farmers. School of American Research Press, Santa Fe, NM, 1995, p. 3.
Pringle, H. The slow birth of agriculture. Science 282(5393): Nov. 20, 1998.
Pyler, E. J. Our Daily Bread. Siebel, Chicago c1958.
Qarooni, J., Ponte, J. G. and Posner, E. S. Flat breads of the world. Cereal Foods World, Dec. 1992.
Sokal, R., Oden, N. and Wilson, C. Nature, May 1991.
Sokolov, R. The unknown bioengineers. Natural History 102(10): Oct. 1993.
Storck, J. and Teague, W. D. Flour for Man's Bread. University of Minnesota Press, Minneapolis, 1952.
Waehren, M. Le monde du pain il y a environ 5500–5800 ans, Le boulanger-pâtissier. La Press Corporative Française, Paris, LII, 1 January 1983.
Dr. E. Wood in National Geographic, Jan. 1995.
Ziehr, W. Bread through the ages. Lannoo, Tielt, Belgium, 1987.
Zohary, D. and Hopf, M. Domestication of plants in the Old World, 2nd ed. Clarendon Press, Oxford, 1994.

2

Biological Fundamentals of Yeast and Lactobacilli Fermentation in Bread Dough

Peter Stolz
Consultant, Minden, Germany

I. INTRODUCTION

Cereal-based foods play an important role as a source of dietary energy and nutrients in human nutrition. The use of microorganisms by humans has a long tradition. Besides beer and wine production, breadmaking is one of the oldest arts known to man. For example, Egypt, Greece and Italy were early places of discovery of leavened breads, though the existence of microorganisms was not known at that time. However, today we know that lactic acid bacteria (LAB) and yeasts are prevailing organisms in carbohydrate-rich environments. They break down carbohydrates to mainly lactic acid, ethanol/acetic acid, ethanol, and CO_2 respectively. In Germany and in northern and eastern parts of Europe sourdough still plays an important role in the production of rye and rye-mixed breads, despite the increased use of baker's yeast after their introduction in the late 19th century. In southern and western parts of Europe, in North America, and in the Middle East as well as in the typical rye sourdough countries, the application of sourdough for the production of wheat breads is developing. The increased interest in sourdoughs is a result of consumers' demand for (1) bread without additives, (2) flavor-intensive breads, (3) breads with good keeping qualities and shelf life. Moreover, the use of sourdough starter cultures and the development of modern sourdough equipment and processes led to an easier handling in the production of sourdoughs in artisanal and industrial bakeries.

II. FUNCTION OF SOURDOUGH FERMENTATION

Sourdough fermentation has different functions depending on the type of bread. The baking properties of rye and wheat are different from each other. The crumb structure of a wheat bread is based on a gluten matrix, formed by the proteins gliadin and glutenin during the mixing. The gluten matrix is of great importance for the gas retention capacity of a wheat dough. In contrast the crumb structure of rye bread is based on a starch matrix because rye flour is lacking the ability to form gluten. Since rye flour normally has a higher amylase activity than wheat flour, a bread dough containing more than 20% rye flour has to be soured to inhibit the starch degradation by the α-amylase activity. Moreover, the gelatinization temperature of rye starch lies in the same temperature range (rye starch 53–64 °C/wheat starch 63–79 °C) where α-amylase has its optimal activity. Therefore, dough acidification is essential to obtain a proper crumb structure in a rye or rye-mixed bread. In addition, some more effects are obtained when sourdoughs are used in the breadmaking process. The most important effects of a sourdough fermentation for rye and rye-mixed breads are depicted in Table 1.

In general, the more rye flour is used in a bread recipe the more sourdough has to be used in the bread dough. In Germany many guidelines exist that show how bakers should work with sourdough and how much sourdough should be used for different types of bread. These guidelines were summarized by Spicher and Stephan (1993). In Table 2, the general effects of rye and wheat sourdoughs in breads are depicted.

Table 1 Specific Functions of Sourdough for Rye and Rye-Mixed Breads

Safeguarding of the bakability of the rye flour
- Suppresion of rye amylase activity by lowering the pH value (the gelatinization temperature of rye starch lies in the same temperature range, rye starch 53–64 °C/wheat starch 63–79 °C, were the α-amylase has optimal activity)
- formation of the starch matrix of the rye and rye-mixed bread

Increasing water binding capacity of the rye flour by
- Rye pentosans and
- Other swelling substances

Table 2 General Functions of Sourdough for Rye and Wheat Breads

Formation of sourdough flavor by
- Aroma precursors and aroma substances e.g., peptides, amino acids, reducing sugars, lactate, acetic acid, ethanol, other alcohols, esters, carbonyls, etc.
- Antimicrobial activity by lowered pH value and acetic acid
- Mold inhibition
- Inhibition of rope-causing microorganisms

Improved dough properties by
- Higher water binding capacity, succulent crumb
- Improved dough elasticity
- Improved machinability

Improved shelflife
- retrogradation retarded
- improved fresh keeping qualities
- Improved crustiness
- Improved color
- Maillard reaction (browning reactions)

III. MICROBIOLOGICAL FUNDAMENTALS OF SOURDOUGH

This chapter presents the microbiology of sourdough in terms of its microbial–ecological aspects and considers new findings.

Sourdough fermentations are usually carried out under limited air contact. This leads to the formation of microflora that influences the dough by their fermentation products in such a way that only selected acid-tolerant microorganisms can resist the prevailing conditions. Cereal mashes with a pH of 5–6.2, which are rich in fermentable carbohydrates, will be preferentually fermented by LAB, at least to a pH below 4. Below that point, acid-tolerant yeasts dominate the fermentation. Therefore, spontaneous fermentations of carbohydrate-rich and weak acid doughs are usually characterized by the succesive or contemporary growth of LAB and yeasts. It is remarkable that in sourdoughs propagated under different conditions a distinct microflora can be found that is specially adapted to the prevailing process conditions. In some cases, the microflora consists of strains that were never found in habitats other than sourdough (Hammes and Vogel, 1997). Factors affecting the microbial growth and metabolism of the sourdough microflora are shown in Fig. 1.

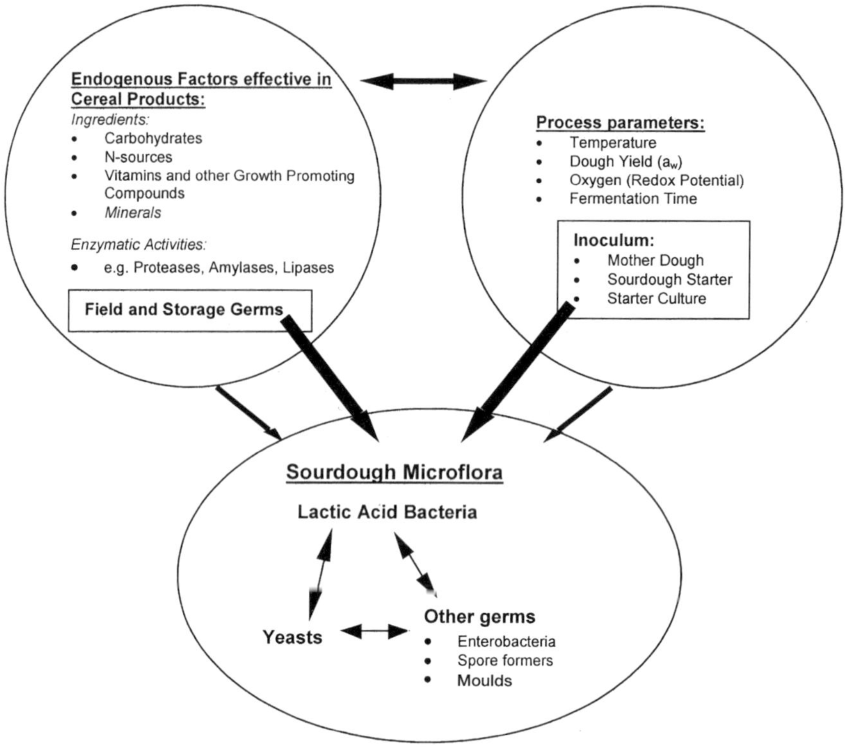

Figure 1 Factors affecting growth and metabolism of sourdough microflora.

Not before the 20th century was the key role of LAB and yeasts in sourdough recognized by Holliger, Beccard, and Knudsen. Beccard (1921) and Knudsen (1924) were working on the development of growth media. They already knew that LAB have only limited biosynthetic properties and have high requirements on the availability of amino acids, peptides, purines, and pyrimidines. Even today one cannot be sure that the complete microflora of a sourdough can be isolated and determined. In future, new molecular biological techniques will determine whether all microorganisms can be grasped and cultivated on agar plates.

A. Microflora of Spontaneously Fermented Doughs

A spontaneous dough fermentation starts by mixing flour with water without adding a starter culture or a portion of a preceding sourdough (mother dough). The microflora of such a dough depends on the microflora of the raw materials used (flour/cereal, water, other ingredients) and the prevailing hygienic conditions, and is variable in terms of kind, origin, and storage conditions of the flour as well as the technological parameters of the fermentation process applied (Fig. 1). In Table 3 bacteria isolated from cereals and flours thereof are listed (Spicher et al., 1997).

B. Bacteria of Spontaneously Fermented Doughs

In the early stages of a spontaneous fermentation, gram-negative enterobacteria (facultative anaerobic rods) will dominate the fermentation. During the course of the fermentation, normally gram-positive LAB gain dominance (Rohrlich, 1957/58; Spicher, 1966; Lönner et al., 1986; Hochstrasser et al., 1993). A variety of LAB and yeasts were isolated from spontaneously fermented doughs. The dominating LAB of already fermented spontaneous sourdoughs are homofermentative lactobacilli and pediococci. These species were found in spontaneous wheat and rye sourdoughs with cell counts between 3×10^8 and 3×10^9 colony-forming units per gram dough (cfu/g) (Spicher and Stephan, 1966; Lönner et al., 1986; Spicher et al., 1987; Galli et al., 1987). In Table 4 an overview of

Table 3 Detectable Bacteria of Cereals and Flours Thereof

Characteristics	Important as
Gram-negative, aerobic rods genera: *Pseudomonas, Xanthomonas, Acinetobacter, Flavobacterium Alcaligenes*	fh
Gram-negative, facultative anaerobic rods genera: *Escherichia, Salmonella, Citrobacter, Klebsiella, Enterobacter, Erwinia, Proteus*, etc.	p/p/tf/ifi
Gram-positive cocci genera: *Micrococcus, Staphylococcus, Pediococcus, Enterococcus, Leuconostoc, Sarcina*	fh/p/tf
Gram-positive rods, endospore-forming genera: *Bacillus, Clostridium*	fh/p/p/tf
Gram-positive rods, regular, non-spore-forming genera: *Lactobacillus, Weissella*	fh
Gram-positive rods, irregular, non-spore-forming genera: *Corynebacterium, Brevibacterium, Propionibacterium, Actinomyces*	p/tf

fh, food-hygienic important species; p, pathogenic species; p/tf, pathogenic and toxin-forming species; ifi, indicator of faecal infections.

Table 4 Homofermentative and Heterofermentative LAB Found in Spontaneous Sourdoughs

Homofermentative LAB	Heterofermentative LAB
Lb. casei	*Lb. brevis*
Lb. delbrueckii	*Lb. buchneri*
Lb. farciminis	*Lb. fermentum*
Lb. plantarum	
Pc. acidilactici	
Pc. pentosaceus	

LAB, Lactic acid bacteria; *Lb., Lactobacillus*; *Pc., Pediococcus.*

homo- and heterofermenative LAB isolated from already fermented spontaneous sourdoughs is presented.

C. Yeasts of Spontaneously Fermented Doughs

Our understanding of the bacteria of spontaneously fermented sourdoughs is much better than that of their yeasts. Heinz and Klaushofer (1952) presented early information about this topic. Based on the morphology and the physiological properties, four different yeast species could be discriminated (*Saccharomyces turbidans, Saccharomyces marchalianus, Torulopsis albida, Torula holmii* syn. *Saccharomyces exiguus*). The latter three species were designated as important since these species were acid tolerant and showed a good gas formation capacity. In spontaneous sourdoughs Spicher and Schöllhammer (1977) found yeasts with cell counts between 2.5×10^6 and 6.7×10^7 cfu/g. The isolated yeasts were identified as *Saccharomyces cerevisiae* and *Pichia saitoi*, according to Beech and Davenport (1971) and Lodder (1970). In 1983, *Saccharomyces minor* and *Saccharomyces cerevisiae* were found by Kazanskaya et al. in spontaneous sourdoughs from Russian bakeries (1983).

D. Sourdough Microflora

In contrast to spontaneously fermented sourdoughs, traditional sourdough processes rely not on a fortuitous flora but on the use of commercial starter cultures or a portion of a preceding sourdough (mother dough, chef, madre, Anstellgut). The range of commercial cultures includes (1) pure starter cultures in powder form, i.e., freeze-dried single-strain or mixed-strain cultures, and (2) starter cultures that are active sourdoughs. The use of pure starter cultures, as is common in the milk, wine, and meat industries, has not yet been successfully used for sourdough production. In contrast sourdough starter cultures, in Germany called "Reinzuchtsauerteig", have been successfully applied for decades. For example, Böcker Reinzuchtsauerteig is a sourdough which has been continuously propagated for several decades and has led to a stable microflora consisting of *Lb. sanfranciscensis* (synonym *Lb. sanfrancisco* and *Lb. brevis* ssp. *lindneri*) and *Candida milleri* (Spicher and Schröder, 1978; Böcker et al., 1990; Trüper and De Clari, 1997). In practice, German bakers normally start their sourdoughs regularly once a week or twice a month with a Reinzuchtsauerteig. The sourdough fermentations are performed as single- or multiple-stage processes. During the week, a mother dough is removed from the active sourdough which is used to start the sourdough fermentation for the next day. The microflora of this Reinzuchtsauerteig dominates such sourdough fermentations and avoids undesirable results, respectively.

Traditionally, propagated sourdoughs are the only food fermentation processes that are dominated by obligate heterofermentative lactobacilli (Kline and Sugihara, 1971; Spicher and Schröder, 1978; Böcker et al., 1990; Ottogalli et al., 1996). The fermentation parameters of traditional one-stage or three-stage sourdough processes ensure (Fig. 1) that the microorganisms remain metabolically active throughout the process. In such sourdoughs, only a few heterofermentative species, especially *Lb. sanfranciscensis* (synonym *Lb. sanfrancisco* and *Lb. brevis* ssp. *lindneri*) and *Lb. pontis* can be found in numbers from 10^8 to 2×10^9 cfu/g, which suggests a relevant contribution to the fermentation process (Hammes et al., 1996; Vogel et al., 1996).

In contrast to homofermentative lactobacilli, these obligate heterofermentative species exhibit metabolic activities that contribute to the competitiveness of these strains in sourdough and are important for technologically relevant properties of lactobacilli in sourdough, i.e., rapid acidification, balanced production of lactic and acetic acid, and gas production. The key reactions of the carbohydrate metabolism of *Lb. sanfranciscensis* and *Lb. pontis* are an effective maltose metabolism via the enzyme maltose phosphorylase and the capability to use fructose and oxygen as electron acceptors (Stolz et al., 1993, 1985a,b, 1996; Stolz, 1995; Hammes et al., 1996).

1. Bacteria of Sourdough

For over 100 years sourdough is a topic of scientific investigations. There exist numerous publications about the microflora of sourdough. In 1894 Wolffin found bacteria and yeasts in sourdough (Rohrlich, 1976). One of the first systematic investigations on the microflora of sourdough were carried out by Holliger (1902), who demonstrated that the sourdough fermentation was accomplished by bacteria which he attributed to the Lactobacillaceae. Henneberg (1903) was the first to report on heterofermentative LAB in sourdough, which he found beside homofermentative LAB and yeasts. Beccard (1921) isolated homo- and heterofermentative lactobacilli from rye sourdoughs used for the production of whole-meal breads and pumpernickel. As a result of his baking trials he mentioned that the gas- and acetic acid–producing heterofermentative lactobacilli are the typical sourdough bacteria. Knudsen (1924) isolated more than 200 strains from sourdoughs of Danish bakeries, which he subdivided in different groups. The strains were classified as *Streptobacterium plantarum* syn. *Lb. plantarum, Thermobacterium cereale* syn. *Lb. delbrüeckii,* and *Betabacterium.* According to differences in the group *Betabacterium,* Knudsen subdivided the group into *Betabacterium* α, β, and γ. The growth of *Betabacterium* γ was poor on the medium first used, but after the addition of a cereal mash and a yeast autolysate *Betabacterium* γ grew well. The optimal growth of these strains was obtained at temperatures between 25 °C and 30 °C. Knudsen (1924) observed that the homofermentative strains form mainly lactic acid and the heterofermentative strains form acetic acid and other byproducts besides lactic acid. He also concluded that *Betabacterium* γ in particular was best adapted to sourdough and that these bacteria were responsible for the typical flavor and aroma of sourdough.

The research activities on sourdough were summarized in reviews and articles from Sugihara (1985), Spicher (1987), Salovaara (1993), Hammes et al. (1996), Hammes and Vogel (1997), and Hammes and Gänzle (1997). In Germany, Spicher et al. worked for more than 25 years on sourdough, which brought about basic knowledge in that field. Spicher isolated numerous lactobacilli and yeasts from spontaneously fermented sourdoughs and sourdough starters called Reinzuchtsauerteig (Spicher, 1959, 1984; Spicher and Stephan, 1966; Spicher and Schröder, 1980; Spicher et al., 1980; Spicher and Lönner, 1985). Spicher's investigations with *Lb. brevis* ssp. *lindneri* yielded best results according to quality parameters (acidifying performance, pore picture, crumb elasticity, aroma, and flavor) of rye-sourdough breads produced with different sourdough processes (three stage, one Stage, and Berliner short sourdough processes). The positive influence of *Lb. brevis* ssp. *lindneri* (synonym *Lb. sanfrancisco* and *Lb. sanfranciscensis*) on bread flavor and texture was also demonstrated by Hansen et al. (1989), Hansen and Hansen (1996), and Gobbetti et al. (1996).

In California wheat sourdoughs are used for the production of San Francisco Sourdough Bread. Kline and Sugihara (1971) found three different strains of a heterofermentative lactobacilli and the yeast *Torulopsis holmii.* This microbial association was reported to be characteristic and stable in these traditional prepared wheat sourdoughs that were also continuously propagated over a long time. For the heterofermentative lactobacilli, Kline and Sugihara (1971) suggested the name *Lb. sanfrancisco* in honor of their unique role for this process.

Ottogalli et al. (1996) confirmed these results with his investigations on Italian wheat sourdoughs for the production of panettone. They also found *Lb. sanfrancisco* and *Candida holmii* as the dominating microflora, both in ratios of 100:1.

In 1984, Weiss and Schillinger demonstrated with DNA-DNA homology studies that *Lb. brevis* ssp. *lindneri* strains found by Spicher in rye sourdoughs (Reinzuchtsauerteige) could be allotted to the species *Lb. sanfrancisco* found by Kline and Sugihara (1971) in wheat sourdoughs (Kandler und Weiss, 1986). In accordance with rule 12c of the International Code of Nomenclature for bacteria, the name *Lb. sanfrancisco* was corrected to *Lb. sanfranciscensis* (Trüper and De Clari, 1997).

Sourdough bacteria cannot be seen as an independent group of bacteria existing only in sourdough. Rather they can be described as a group of specially adapted varieties of LAB that are also common in other habitats. The most relevant bacteria isolated from sourdough are listed in Table 5 and belong to the "genus" *Lactobacillus*. Up to now, *Lb. sanfranciscensis* (syn. *Lb. sanfrancisco, Lb. brevis* ssp. *lindneri*) and *Lb. pontis* have been the only species isolated from sourdough exclusively.

Characteristics of lactic acid bacteria (LAB). A first attempt to classify LAB was made by Orla-Jensen (1919). Based on the differences in growth temperature and morphology, he discriminated between *Thermobacterium, Streptobacterium*, and *Betabacterium*. Kandler and Weiss (1986) arranged LAB on the basis of biochemical-physiological criteria because many newly described strains did not fit into the traditional classification scheme of Orla-Jensen. The actual list of genera belonging to the LAB (Lactobacillaceae) is given in Table 6 (Wood and Holzapfel, 1995).

The persistent increase of new species is caused by ongoing investigations of habitats and the development of new analytical methods, i.e., "fingerprint" analysis of whole-cell proteins with sodium dodecyl sulfatepolyacrylamide gel electrophoresis, DNA or RNA sequence and hybridization methods, as well as RAPD and other polymerase chain reaction (PCR)–based methods, which allow a more precise identification and taxonomic classification of the isolated bacteria. As a consequence of 16S rRNA studies, the genus *Lactobacillus* was regrouped into three groups (Collins et al., 1991): a *Lb. delbrueckii* group, a *Lb. casei*/*Pediococcus* group, and a *Leuconostoc* group. An excellent review of the history of the classification of LAB was given by Pot et al. (1994). In the newest attempt to

Table 5 List of Lactic Acid Bacteria Isolated from Sourdough and Their Classification in Between the Genus *Lactobacillus*

Group[a]	Lactobacillus
Aa	*Lb. acidophilus, Lb. amylovorus, Lb. crispatus, Lb. delbrueckii (Lb. delbrueckii* ssp. *bulgaricus, Lb. delbrueckii* ssp. *delbrueckii, Lb. delbrueckii* ssp. *lactis), Lb. johnsonii*
Ab	*Lb. farciminis*
Bb	*Lb. alimentarius, Lb. casei, Lb. plantarum*
Cb	*Lb. brevis, Lb. buchneri, Lb. fermentum, Lb. fructivorans, Lb. reuteri, Lb. pontis, Lb. sanfranciscensis*
Cc	*Lb. confusus*[b] (now *Weissella confusus*)
	Weissella
	Weissella confusus
	Pediococcus
	Pediococcus acidilactici, Pediococcus pentosaceus

[a] Actual classification of lactobacilli.
[b] The species *Lactobacillus confusus* is now alloted to the genus *Weissella* (Collins et al., 1993).

classify the genus *Lactobacillus* (Hammes and Vogel, 1995), the investigators tried to combine the biochemical-physiological grouping criteria of Kandler and Weiss (1986) (A, B, C) with the phylogenetic relationship of Collins et al. (1993) (a, b, c) and to consider the murine arrangement of the bacteria–cell wall (Fig. 2). At the moment, 65 species are published and adopted in the "approved list" of Skerman et al. (1980). In the meantime, 9 species have been transfered to other genera or species leaving 56 species, which can readily be allotted to three evolutionary related groups: *Lb. delbrueckii, Lb. casei/Pediococcus*, and *Leuconostoc*. The lactobacilli is the genus with the

Table 6 Genera Belonging to the Lactic Acid Bacteria (Lactobacillaceae)

Bifidobacterium	Streptococcus
Carnobacterium	Tetragenococcus
Enterococcus	Vagococcus
Lactobacillus	Weissella
Lactococcus	Aerococcus
Leuconostoc	Alloiococcus
Pediococcus	Atopobium

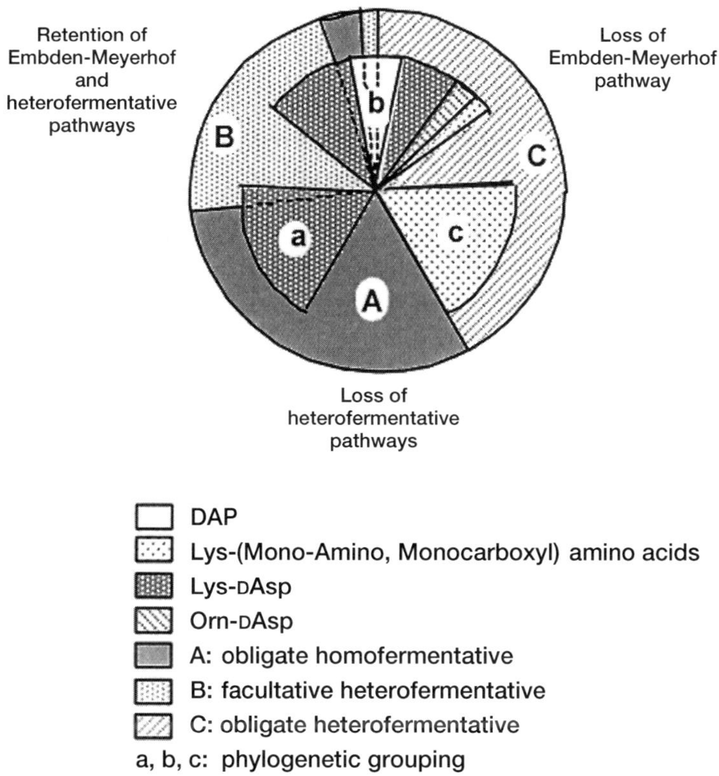

Figure 2 Grouping of lactobacilli.

most species in the LAB. Hereinafter the actual classification of the genus *Lactobacillus* is described (Hammes and Vogel, 1995):

Group A: Obligate homofermentative lactobacilli. Hexoses are fermented nearly completely to lactic acid (>85%) via the Embden-Meyerhof-Parnas (EMP) pathway. The organisms possess the enzyme fructose-1,6-biphosphate aldolase but no phosphoketolase. Therefore, they are not able to ferment gluconic acid and pentoses.

Group B: Facultative heterofermentative lactobacilli. The organisms possess the enzyme fructose-1,6-biphosphate aldolase plus a phosphoketolase; therefore, they are able to ferment hexoses and pentoses (gluconic acid as well). They ferment hexoses nearly completely to lactic acid via the EMP. Pentoses are fermented via the phosphogluconate pathway to lactic acid and acetic acid. In the presence of glucose, the enzymes of the phosphogluconate pathway are inhibited.

Group C: Obligate heterofermentative lactobacilli. The organisms possess the enzyme phosphoketolase. They ferment hexoses to lactic acid, ethanol or acetic acid, and CO_2 in equimolar amounts via the phosphogluconate pathway. Pentoses are also fermented via the phosphogluconate pathway to lactic acid and acetic acid.

Within the three groups *A, B*, and *C*, the species are grouped according to their phylogenetic membership. The letter *a* indicates the membership to the *Lb. delbrueckii* group, *b* to the *Lb. casei/Pediococcus* group, *c* to the *Leuconostoc* group. For example, the combination of the letters *Aa* indicates the species as an obligate homofermentative member of the *Lb. delbrueckii* group, whereas *Cb* would indicate that a species is obligate heterofermentative and belongs to the *Lb. casei/Pediococcus* group (Fig. 2).

2. Yeasts of Sourdough

For the taxonomic characterization of yeasts, not long ago only physiological and microscopic methods were used. These methods use the capabilities of the yeasts to ferment sugars anaerobically, to use them for growth under aerobic conditions, to use nitrate to form ascospores, myceles, or pseudomyceles, respectively (Spicher et al., 1979). Nowadays the application of modern methods such as ribotyping or PCR techniques will probably also result in dramatic changes for the yeast systematics as is true for the taxonomy of the bacteria.

The first indications of yeasts from sourdough were given by Peters (1889), Boutroux (1892), and Wolffin (1894). Budinoff (1903) found *Saccharomyces cerevisiae* in rye bread doughs. Henneberg (1909) reported about different yeasts from sourdoughs, and Maurizio (1917) found *Saccharomyces minor* in sourdoughs from German bakeries in Würzburg. In the years 1939–1955, Pelshenke, Schulz, Rohrlich, and Pokorny described different types of yeasts isolated form sourdoughs (Pelshenke et al., 1939; Pelshenke and Schulz, 1942; Pokorny, 1955). In 1965 Trojan isolated a maltose-negative yeast, which he described as *Torulopsis holmii*.

In the early 1970s, Kline and Sugihara investigated the microflora of a traditional wheat sourdough used for the production of San Francisco sourdough French bread (Kline et al., 1970; Kline and Sugihara, 1971; Sugihara et al., 1970, 1971). Their work revealed that the predominant yeast was *Torulopsis holmii*, a maltose-negative yeast, which is the imperfect form (nonsporulating) of *Saccharomyces exiguus*. These authors assumed that the stability of the microflora system, *Lb. sanfrancisco* and *Torulopsis holmii*, results from the inability of the yeast to ferment maltose because there is no competition on maltose, the main carbon source in sourdough.

Yeasts were found in wheat, rye, and in sorghum sourdoughs with amounts of 0.1–10% to the total counts of microorganisms (Sugihara et al., 1971; Spicher et al., 1979; Foramitti and Mar, 1982; Spicher and Stephan, 1987; Hamad et al., 1992; Böcker, 1993). Spicher et al. (1979) isolated *Saccharomyces cerevisiae, Pichia saitoi, Candida krusei*, and *Torulopsis holmii* from two different commercial sourdough starter cultures in Germany. Salovaara and Savolainen (1984) found *Saccharomyces cerevisiae* and *Torulopsis holmii* in sourdoughs from Finnish bakeries with cell counts of 5×10^5 to 5×10^8 yeasts/g sourdough. From Finnish households they isolated *Torulopsis unisporus, Torulopsis stellata, Endomycopsis fibuliger*, and *Hansenula anomala* with cell counts of 10^4–10^8 yeasts/g sourdough, but the latter two yeasts were designated as contaminations and were not considered as typical sourdough yeasts. Barber et al. (1983) isolated 2.2×10^4 to 1.3×10^8 yeasts/g from Spanish wheat sourdoughs. The yeasts belonged to the species *Saccharomyces cerevisiae, Pichia polymorpha, Trichosporon margaritiferum, Saccharomyces fructuum*, and *Hansenula subpelliculosa*.

Nout and Creemers-Molenaar (1987) tested the microbiological stability of two Dutch wheat sourdough starters consisting of *Lb. sanfrancisco* (syn. *Lb. sanfranciscensis*) and *Saccharomyces exiguus*, against the ordinary baker's yeast *Saccharomyces cerevisiae*, which exerts great infection pressure on this system. The study revealed that *Saccharomyces cerevisiae* had disappeared after two consecutive fermentations, although *Saccharomyces cerevisiae* was precultivated on maltose. In another study, these authors could demonstrate that an initial sourdough population consisting of *Lb. plantarum, Lb. brevis, Lb. buchneri, Lb. cellobiosus*, and the yeast *Torulaspora delbrueckii* was reduced to a microflora consisting of *Lb. sanfrancisco* and *Torulaspora delbrueckii* after 7 weeks or 20 dough refreshments, respectively, though *Lb. sanfrancisco* was not assignable at the beginning of the investigation.

Böcker et al. (1990) confirmed the results of Nout and Creemers-Molenaar. They investigated a commercial sourdough starter culture (Reinzuchtsauerteig) produced with a traditional sourdough method following the principle of consecutive cycles from the previous batch. Over a period of 10 months they found a stable microflora consisting of *Lb. sanfrancisco* and *Candida milleri*. Gänzle et al. (1998) investigated the effect of ecological factors (i.e., pH, temperature, ionic strength, lactate, acetate, and ethanol) and set up a mathematical model that describes the effects of these parameters on the growth of *Candida milleri* and two strains of *Lb. sanfranciscensis*. They observed that these organisms exhibited the same response in a temperature range around 26 °C, where sourdough fermentations are commonly performed (24 °C and 28 °C). Moreover, low pH values and acetate concentrations of up to 80 mmol/L had little effect on the growth of *C. milleri*. These findings provide insight into factors contributing to the stability of a sourdough microflora.

The microflora of traditionally prepared wheat sourdoughs for the production of panettone (Ottogalli et al., 1996) consisted of *Lb. sanfranciscensis* and *Candida holmii*, the unsexual form of *Saccharomyces exiguus*. *Candida holmii* was found with 10^6–10^7 cfu/g and *Lb. sanfranciscensis* with 10^8–10^9 cfu/g at the beginning and end of the fermentation. The ratio of yeasts to lactobacilli was always around 1:100.

The same authors found a heterogeneous microflora in wheat sourdoughs for the production of Puglieser bread, consisting of *lactobacilli* (*Lb. plantarum, Lb. brevis, Lb. fermentum*, and *Lb. fructivorans*) with 10^7–10^9 cfu/g, lactic acid–forming cocci (*Streptococcus, Pediococcus*, and *Leuconostoc*) with 10^5–10^9 cfu/g and a yeast flora, mainly consisting of *Saccharomyces cerevisiae* which was added intentionally and therefore

Table 7 Yeasts Isolated from Sourdoughs

A	B
Candida boidinii	*Saccharomyces exiguus*[a]
Candida guilliermondii	*Candida holmii*[b]
Candida stellata	*Candida krusei*[c]
Candida tropicalis	*Candida milleri*[d]
Hansenula anomala	*Pichia saitoi*
Hansenula subpelliculosa	*Torulopsis holmii*[e]
Hansenula tropicalis	
Pichia polymorpha	
Saccharomyces cerevisae	
Saccharomyces dairensis	
Saccharomyces ellipsoideus	
Saccharomyces fructuum	
Saccharomyces inusitatus	

Columns A and B: yeasts isolated from wheat and rye sourdoughs; column B: typical sourdough yeasts.

[a] Formerly *Torulaspora delbrueckii*.

[b] Asexual form from *Torulopsis holmii* var. *holmii* syn. *Saccharomyces exiguus*.

[c] Asexual form from *Issatchenkia orientalis*.

[d] Asexual form from *Torulopsis holmii* var. *acidi lactici* syn. *Saccharomyces exiguus*.

[e] Syn. *Saccharomyces exiguus* and sexual form from *Candida holmii*.

cannot be seen as a natural component of this sourdough microflora. In this case, the ratio of yeasts to LAB was about 1:10.

In columns A and B of Table 7 a list of yeasts found in sourdoughs is enumerated. Column B lists all yeasts, which can been designated as typical sourdough according to current knowledge.

REFERENCES

Barber, S., R. Baguena, M. A. Martinez-Anaya and M. J. Torner (1983). Microflora de la masa madre panaria. I. Identificacion y propiedades funcionales de microorganismos de masas madre industriales, elaboradas con harina de trigo. Rev. Agroquim. Tecnol. Alimentos 23(4): 552–562.

Beccard, E. (1921). Beiträge zur Kenntnis der Sauerteiggärung. Zbl. f. Bakt. Il Abt. 112, **54**(20/24): 465–471.

Beech, F. W. and R. R. Davenport (1971). Isolation, purification and maintenance of yeasts. Meth. Microbiol. **4**: 153–182.

Böcker, G., R. F., Vogel and W. P., Hammes (1990). *Lactobacillus sanfrancisco* als stabiles Element in einem Reinzucht-Sauerteig-Präparat. Getreide Mehl Brot **44**: 269–274.

Böcker, G. (1993). Ökologische und physiologische Charakterisierung der sauerteigtypischen Stämme *Lactobacillus sanfrancisco* und *Lactobacillus pontis* sp. nov. Dissertation, Institut für Lebensmitteltechnologie, Universität Hohenheim.

Boutroux, L (1892). Le pain et la panification. Ann. Chim. Phys. **6**: ser. T 26, 1.

Budinoff, L. (1903). Die Mikroorganismen der Schwarzbrotgärung. Zbl. f. Bakt. Il. Abt. **10**: 458–463.

Collins, M. D., U. Rodrigues, C. Ash, M. Aguirre, J. A. E. Farrow, A. Martinez-Murcia, B. A. Phillips, A. M. Williams and S. Wallbanks (1991). Phylogenetic analysis of the genus *Lactobacillus* and related lactic acid bacteria as determined by reverse transcriptase sequencing of 16S rRNA. FEMS Microbiol. Lett. **77**: 5–12.

Collins, M. D., J. Samelis, J. Metaxopoulos and S. Wallbanks (1993). Taxonomic studies on some leuconostoc-like organisms from fermented sausages: description of a new genus *Weissella* for the *Leuconostoc paramesenteroides* group of species. J. Appl. Bacteriol. **75**: 595–603.

Foramitti, A. and A. Mar (1982). Neue Aspekte und Erkenntnisse zur kontinuierlichen Herstellung von Sauerteig in der Brotindustrie. Getreide Mehl Brot **36**: 147–149.

Gänzle, M. G., M. Ehmann, W. P. Hammes (1998). Modelling of growth of *Lactobacillus sanfranciscensis* and *Candida milleri* in response of process parameters of the sourdough fermentation. Appl. Environ. Microbiol.

Galli, A., L. Franzetti and M. G. Fortina (1987). Isolation and identification of yeasts and lactic acid bacteria in wheat flour. Microbiol. Alim. Nutr. **5**: 3–9.

Gobetti, M., E. Smacchi and A. Corsetti (1996). The proteolytic system of *Lactobacillus sanfrancisco* CB1: purification and characterization of a proteinase, a dipeptidase and an aminopeptidase. Appl. Environ. Microbiol. 62(9): 3220–3226.

Hamad, S. H., G. Böcker, R. F. Vogel and W. P. Hammes (1992). Microbiological and chemical analysis of fermented sorghum dough for Kisra production. Appl. Microbiol. Biotechnol. **37**: 728–731.

Hammes, W. P. and R. F. Vogel (1995). The genus *Lactobacillus*. In: Wood, B. J. B., and W. Holzapfel (eds.). The Genera of Lactic Acid Bacteria, Vol. 2. Blackie, London, pp. 19–54.

Hammes, W. P., P. Stolz and M. Gänzle (1996). Metabolism of lactobacilli in traditional sourdough. Arch. Food Sci. **18**: 176–184.

Hammes, W. P. and R. F. Vogel (1997). Sauerteig. In: Müller, G., W. Holzapfel and H. Weber (eds.). Mikrobiologie der Lebensmittel: Lebensmittel pflanzlicher Herkunft. Behr's Verlag, Hamburg, pp 263–285.

Hammes, W. P. and M. G. Gänzle (1997). Sourdough breads and related products. In: Wood, B. J. B. (ed.). Microbiology of Fermented Foods. Chapman and Hall, London, pp. 199–216.

Hansen, Å. B. Lund and M. J. Lewis (1989). Flavour of sourdough rye bread crumb. Lebensm. Wiss. Technol. **22**: 141–144.

Hansen, A, B. Hansen (1996). Flavour of sourdough wheat bread crumb. Z. Lebensm. Untersuch. Forsch. **202**: 244–249.

Heinz, L. and H. Klaushofer (1952). Studien über die Flora des Sauerteiges. Mitteilungen Gärungsgewerbe Wien **9/10**: 1–7.

Henneberg, W. (1903). Zur Kenntnis der Milchsäurebakterien der Brennereimaische, der Milch, des Bieres, der Presshefe, der Melasse, des Sauerkohls, der sauren Gurken und des Sauerteiges, sowie einige Bemerkungen über die Milchsäurebakterien des menschlichen Magens, Z. Spiritusind., **26**, 22–31; sec: Zentralblatt für Bakteriologie, Parasitenkunde, Infektionskrankheiten und Hygiene, **11**, 163.

Henneberg, W. (1909). Gärungsbakteriologisches Praktikum, Betriebsuntersuchungen und Pilzkunde. Paul Parey Verlag, Berlin, pp. 498–499.

Hochstrasser, R. E., A. Ehret, O. Geiges and W. Schmidt-Lorenz (1993). Mikrobiologie der Brotteigherstellung. Il. Mikrobiologische Untersuchunen an unterschiedlich hergestellten Spontansauerteigen aus Weizenmehl. Mitt. Gebiete Lebensm. Hyg. **84**: 357–381.

Holliger, W. (1902). Bakteriologische Untersuchungen über Mehlteiggärung. Zbl. f. Bakt. II 9: 305–312, 361–371, 395–425, 473–483, 521–537.

Kandler, O. and N. Weiss (1986). Regular, non-sporing gram-positive rods. In: Sneath, P. H. A., N. S. Mair, M. E. Sharpe and J. G. Holt (eds.). Bergey's Manual of Systematic Bacteriology, Vol. 2. William and Wilkins, Baltimore.

Kazanskaya, L. N., O. V. Afanasyeva and V. A. Patt (1983). Mikroflora of rye sours and some specific features of its accumulation in bread baking plants of USSR. Proc. In: Progress in Cereal Chemistry and Technology, Developments in Food Science 5B. (J. Holas and Kratochvil, eds.). Elsevier Science, Amsterdam, pp. 759–765.

Kline, L., T. F. Sugihara and L. B. McCready (1970). Nature of the San Francisco sour dough French bread process. I. Mechanics of the process. Bakers Digest **44**: 48–50.

Kline, L. and T. F. Sugihara (1971). Microorganisms of the San Francisco sour dough bread process. II. Isolation and characterization of undescribed bacterial species responsible for the souring activity. Appl. Microbiol. **21**: 459–465.

Knudsen, S. (1924). Über die Milchsäurebakterien des Sauerteiges und ihre Bedeutung für die Sauerteiggärung. Dissertation: Den. Kgl. Veterinaer-og Landbohoskoles Arsskrift.

Lodder, J. (1970). (Ed.) The Yeasts. A Taxonomic Study, 2nd ed. North-Holland, Amsterdam.

Lönner, C., T. Welander, N. Molin, M. Dostálek and E. Blickstad (1986). The microflora in a sour dough started spontaneously on typical Swedish rye meal. Food Microbiol. **3**: 3–12.

Maurizio, A. (1917). Die Nahrungsmittel aus Getreide. Vol. I, Paul Parey, Berlin.

Neumann-Pelshenke (1954). Brotgetreide und Brot. Verlag Paul Parey, Berlin.

Nout, M. J. R. and T. Creemers-Molenaar (1987). Microbiological properties of some wheatmeal sourdough starters. Chem. Mikrobiol. Technol. Lebensm. **10**: 162–167.

Orla-Jensen, S. (1919). The Lactic Acid Bacteria. A. F. Hst and Son. Koeniglicher Hof-Boghandel, Copenhagen.

Ottogalli, G., A. Galli and A. Foschino (1996). Italian bakery products attained with sourdough: characterization of the typical sourdough flora. Adv. Food Sci. (CMTL), **18**: 131–144.

Pelshenke, P. F., E. Weimershaus and J. Svenson (1939). Mikrobiologische Sauerteigstudien. Vorratspflege and Lebensm. Forsch. **II**: 591–603.

Pelshenke, P. F., A. Schulz (1942). Untersuchungen über Sauerteighefe. Vorratspflege Lebens. Forsch. **5**: 154–163.

Peters, W. L. (1889). Die Organismen des Sauerteiges und ihre Bedeutung für die Brotgärung. Bot. Zeitung. **47**, 405.

Pokorny, K. O. (1955). Mikroflora Zitnych Kvasu. Prumsyl potravin **6**: 177–179.

Pot, B, W. Ludwig, K. Kersters and K. H. Schleifer (1994). Taxonomy of lactic acid bacteria. In: Bacteriocins of Lactic Acid Bacteria: Genetics and Applications. (L. De Vuyst and E. J. Vandamme, eds.). Chapman and Hall, Glasgow, pp. 13–89.

Rohrlich, M. (1957/58). Untersuchungen über den Einfluss von Alkoholzusätzen auf die Entwicklung der sog. Spontangärung von Roggenmehlteigen sowie über die Beziehung zwischen Wachstum der Hefen und der Bakterienflora. Jahresbericht Bundesforschungsanstalt für Getreideverarbeitung, 52–55.

Rohrlich, M. (1976). Der Sauerteig historisch betrachtet. Brotindustrie **1**: 9–14.

Rohrlich, M. and W. Essner (1952). Untersuchungen zur Biochemie der Sauerteiggärung. Brot Gebäck **6**(8): 116–120.

Rohrlich, M. and E. Timm (1952/53). Jahresbericht der Bundesforschungsanstalt für Getreideverwertung, Berlin, 24.

Salovaara, H. and J. Savolainen (1984). Yeast type isolated from finnish sour rye dough starters. Acta Alimen. Pol. **10**: 241–245.

Salovaara, H. (1993) Lactic acid bacteria in cereal-based products. In: S. Salminen and A. von Wright (eds.). Lactic acid Bacteria. Marcel Dekker, New York, pp. 111–126.

Skerman, V. B. D., V. McGowan and P. H. A. Sneath (1980). Approved list of bacterial names. Int. J. Syst. Bateriol. **30**: 225–420.

Spicher, G. (1959). I. Mitteilung: Untersuchungen über die Art der in Sauerteigen anzutreffenden stäbchenförmigen Milchsäurebakterien (Genus *Lactobacillus* Beijerinck). Zbl. f. Bakt. II. Abt., **113**: 80–106.

Spicher, G. (1966). Ergebnisse neuerer mikrobieller Sauerteigstudien. Deutsche Lebensm. Rundschau **62**: 43–46, 78–81.

Spicher, G. and H. Stephan (1966). Die Mikroflora des Sauerteiges. III. Untersuchungen über die Art der in "Spontansauern" anzutreffenden Milchsäurebakterien und ihre backtechnische Bedeutung. Zentr. Bakt. Parasitenkunde Infect. Hygiene II. Abt. 120: 699–701.

Spicher, G. and K. Schöllhammer (1977). Vergleichende Untersuchungen über die in Reinzuchtsauerund Spontansauerteigen anzutreffenden Hefen. Getreide Mehl Brot **31**: 215–222.

Spicher, G. and R, Schröder (1978). Die Mikroflora des Sauerteiges. IV. Untersuchung über die Art der in "Reinzuchtsauern" und in Sauerteigen anzutreffenden stäbchenförmigen Milchsäurebakterien (Genus *Lactobacillus* Beijerinck). Z. Lebensm. Untersuch. Forsch. **167**: 342–354.

Spicher, G., R. Schröder and K. Schöllhammer (1979). Die Mikroflora des Sauerteiges. VII. Untersuchungen über die Art der in "Reinzuchtsauern" auftretenden Hefen. Z. Lebensm. Untersuch. Forsch. **169**: 77–81.

Spicher, G. and R. Schröder (1980). Die Mikroflora des Sauerteiges. VIII. Die Faktoren des Wachstums der im "Reinzuchtsauer" auftretenden Hefen. Z. Lebensm. Untersuchung. Forschung **170**, 119–123.

Spicher, G., R. Schröder and H. Stephan (1980). Die Mikroflora des Sauerteiges. X. Die backtechnische Wirkung der in "Reinzuchtsauern" auftretenden Milchsäurebakterien (Genus *Lactobacillus* Beijerinck). Z. Lebensm. Untersuch. Forsch. **171**: 119–124.

Spicher, G (1984). Die Mikroflora des Sauerteiges. XVII. Weitere Untrersuchungen über die Zusammensetzung und die Varabilität der Mikroflora handelsüblicher Sauerteig-Starter. Z. Lebensm. Untersuch. Forsch. **178**: 106–109.

Spicher, G. and C. Lönner (1985). Die Mikroflora des Sauerteiges. XXI. Die in Sauerteigen schwedischer Bäckereien vorkommenden Laktobazillen. Z. Lebensm. Untersuch. Forsch. **181**: 9–13.

Spicher, G., E. Rabe and C. Rohschenkel (1987). Untersuchungen zur Charakterisierung und Bewertung verschiedener Verfahren zur Bereitung eines Spontansauers. 1. Mitteilung: Vergleich verschiedener Spontansauerverfahren. Getreide Mehl Brot **41**: 118–122.

Spicher, G. (1987). Die Mikroflora des Sauerteiges. XXII. Die in Weizensauerteigen vorkommend Laktobazillen. Z. Lebensm. Untersuch. Forsch. **184**: 300–303.

Spicher, G. and H. Stephan (1993) (eds.). Handbuch Sauerteig: Biologie, Biochemie, Technologie Verlag Hamburg, Berlin.

Spicher, G., W. Röcken and H. L. Schmidt (1997).). Getreide und Getreideerzeugnisse. In: Müller, W. Holzapfel and H. Weber (eds.). Mikrobiologie der Lebensmittel: Lebensmittel pflanzlicher Herkunft, Behr's Verlag, Hamburg, pp. 231–260.

Stolz, P., G. Böcker, R. F. Vogel and W. P. Hammes (1993). Utilisation of maltose and glucose by lactobacilli isolated from sourdough. FEMS Microbiol. Lett. **109**: 237–242.

Stolz, P. (1995). Untersuchungen des Maltosemetabolismus von Laktobazillen aus Sauerteig. Dissertation, Institut für Lebensmitteltechnologie, Universität Hohenheim.

Stolz, P., G. Böcker, W. P. Hammes and R. F. Vogel (1995a). Utilization of electron acceptors by lactobacilli isolated from sourdough: I. *Lactobacillus sanfrancisco*. Z. Lebensm. Untersuch. Forsch. **201**: 91–96.

Stolz, P., R. F. Vogel and W. P. Hammes (1995b). Utilization of electron acceptors by lactobacilli isolate from sourdough: II. Lactobacillus pontis, L. amylovorus, L. fermentum and L. reuteri. Z. Lebensm. Untersuch. Forsch. **201**, 402–410.

Stolz, P., W. P. Hammes and R. F. Vogel(1996). Maltose-phosphorylase and hexokinase activity in lactobacilli from traditionally prepared sourdoughs. Adv. Food Sci. (CMTL) **1/2**: 1–6.

Sugihara, T. F., L. Kline and L. B. McCready (1970). Nature of the San Francisco sour dough French bread process. II Microbiological aspects. Bakers Digest **44**: 51–57.

Sugihara, T. F., L. Kline and M. W. Miller (1971). Microorganisms of the San Francisco sourdough bread process. I. Yeasts responsible for the leavening action. Appl. Microbiol. **21**: 456–458.

Sugihara, T. F. (1985). Microbiology of Breadmaking. In: J.B. Wood (ed.), Microbiology of Fermented Foods, Vol. 1. Elsevier Science, London, pp. 249–261.

Trojan, M. (1965). Mikrobiologie der Brotherstellung. Bericht: 2. Tagung Internationale Probleme der modernen Getreideverarbeitung und Getreidechemie, 172–181.

Trüper H. G. and L. De Clari (1997). Taxonaomic note: Necessary correction of specific epithets formed as substantives (nouns) "in apposition." Int. J. Syst. Bacteriol. **47**: 908–909.

Vogel, R. F., M. Müller, P. Stolz and M. Ehrmann (1996). Ecology in sourdoughs produced by traditional and modern technologies. Adv. Food Sci. **18**: 152–159.

Weiss, N. and U. Schillinger (1984). *Lactobacillus sanfrancisco* sp. nov., nom. rev.. Syst. Appl. Microbiol. **5**: 230–232.

Wolffin, A. (1894). Bakteriologische und chemische Studien über die Sauerteiggärung. Med. Diss. Würzburg. Arch. Hyg. **21**: 268.

Wood, B. J. B. and W. H. Holzapfel (eds.) (1995). The Genera of Lactic Acid Bacteria, Vol. 2. Blackie, London.

3
Yeast Fermentations

Daniel H. Maloney* and James J. Foy†
Red Star Yeast and Products, Milwaukee, Wisconsin, U.S.A.

I. CLASSIFICATION

Yeasts are fungi that grow vegetatively as single-celled organisms (see Ref. 47 for a review). The distinction between yeasts and other fungi is sometimes murky; fungi such as rusts and smuts grow as single cells during part of their life cycle and as typical mold-like mycelia during other parts. All yeasts grow vegetatively by budding or by binary fission. Ascomycetous yeasts are distinguished by the fact that during sexual reproduction spores are produced inside a specialized structure known as an *ascus*, whereas basidiomycetous yeasts produce external spores on a structure known as a *basidium*. Although a yeast lifestyle has been adopted by members of three of the four the major fungal groupings (*Basidiomycetes*, *Ascomycetes*, and *Fungi imperfect*, but not *Phycomycetes*), most yeasts of interest to the baking industry are biologically members of the ascomycetous genus *Saccharomyces*. However, because fungal taxonomy relies on the identification of reproductive structures, yeasts that have lost the capacity for sexual reproduction are automatically classified among the *Fungi imperfecti*. Consequently, the San Francisco sourdough yeast, although biologically

---

Current affiliation:
* Medical College of Wisconsin, Milwaukee, Wisconsin, U.S.A.
† Minn-Dak Yeast Sales & Service, Division of Sensient Technologies Corporation, Milwaukee, Wisconsin, U.S.A.

an isolate of *Saccharomyces exiguus*, is also classified variously in the *Fungi imperfecti* as *Torulopsis holmii* or *Candida milleri*.

Yeasts are eukaryotic organisms. Thus, the interior of a yeast cell is divided into compartments with specialized functions: the nucleus, which contains the chromosomes and enzymes required to copy the genetic information needed for protein synthesis; the mitochondrion, where respiration occurs; a vacuole, used for nutrient storage and breakdown of unneeded proteins; membrane vesicles where proteins are modified and targeted to specific cellular locations; and so forth.

Saccharomyces yeasts have been cultivated by humans for thousands of years: wine making dates back to at least 5400–5000 BC [1], and desiccated brewer's yeasts have been found in Egyptian beer residues and bread made about 1000 BC [2]. Over the course of time, *Saccharomyces cerevisiae* has evolved into a species quite different from its modern wild relatives, and has become specialized in the rapid production of carbon dioxide and ethanol from sugars. Its closest relation is probably *S. paradoxus*, a yeast now occasionally isolated from oak sap or soil. Ancient breadmakers probably relied on spontaneous fermentations to leaven their doughs and may have developed sourdoughs at an early date by reserving a portion of a fermented sponge to inoculate the next day's batch. But beer- and wine-making cultures seem to have soon realized that yeast residues from fermented worts and juices would leaven bread as well; until 1850, washed and pressed ale yeasts were still commonly used for this purpose. The popularity of lager beers led to the replacement of brewers' by distillers' yeasts for baking, and since the late 19th century these have been grown specifically for bakers on malt or molasses worts by the commercial yeast producers.

Historically, many different species of *Saccharomyces* yeast were defined on the basis of differences in their ability to ferment carbohydrates. Such criteria were traditionally useful in distinguishing among species of bacteria and fungi that were otherwise very similar. Recent technical advances have made it possible to assign yeast species on the basis of more reliable criteria, such as the ability to visualize yeast chromosomes directly in electrophoretic gels, or DNA sequence comparisons [24]. These analyses confirm the grouping of modern baking, brewing, and wine yeasts into a single species. Just as importantly, they make it possible to identify yeasts belonging to other species. For example, the San Francisco sourdough yeast is clearly a separate species. Most notably, however, modern baking, brewing, wine, and distiller's yeasts all belong to the biological species *S. cerevisiae*. Within this species, differences in carbohydrate utilization depend on minor genetic differences and the yeasts all hybridize readily. Nonetheless, significant differences in fermentation rate, flavor chemicals, and feed preference may be found in different strains.

Table 1 Yeasts Identified in Sourdoughs

Yeast species	Source	Ref.
Saccharomyces cerevisiae	Rye, corn, and wheat sourdoughs	11,12,15,40
S. chevalieri		10
S. curvatus		10
S. exiguus (*T. holmii*, *C. milleri*)	Rye and wheat sourdoughs	10,13,15,19,23,29
S. fructuum	Wheat sourdoughs	16
S. inusitatus	Wheat sourdough (San Francisco)	13
S. panis fermentati		10
Candida boidinii		12,16
C. crusei	Rye and wheat sourdoughs	14,15
C. guillermondii	Wheat sourdoughs	16
C. norvegensis	Wheat sourdough	15
Hansenula anomala	Wheat sourdough	15
H. subpelliculosa	Wheat sourdoughs	16
Pichia satoi	Rye sourdough	14
Torulopsis delbrueckii	Corn and rye sourdoughs	40

Table 1 lists yeast species identified in a sampling of wheat and rye sourdoughs. The list is by no means exhaustive but demonstrates the preponderance of sourdough yeasts from the genus *Saccharomyces*. In fact, of the seven *Saccharomyces* species listed, four (*S. chevalieri, S. curvatus, S. fructuum*, and *S. inusitatus*) are currently considered synonymous with *S. cerevisiae*; one (*S. panis fermentati*) is of uncertain affiliation, and only one, *S. exiguus*, is clearly a separate species [17,18].

II. LIFE CYCLE

The sourdough yeasts identified in Table 1 reproduce vegetatively by budding. In *S. cerevisiae*, bud initiation and the onset of DNA synthesis are roughly simultaneous. As the bud enlarges, DNA synthesis progresses. At the end of synthesis, the nucleus migrates to the neck separating mother and bud, and nuclear division occurs. Unlike animal and plant cell divisions, the nuclear membrane does not break down. Instead, the nucleus itself stretches and divides, with one daughter nucleus migrating into the bud. The bud then separates from the mother cell, leaving a chitin-rich bud scar on the mother cell and a similar birth scar on the bud. In the presence of carbohydrate and protein sources, budding is continuous, and both mother and bud are prepared to begin another division cycle. Industrial yeast strains can complete a division cycle in as little as 75 minutes. An individual cell can produce 20 or so buds before the cell wall becomes covered by bud scars. If nutrients are limited, cells arrest unbudded before initiating another round of DNA synthesis; However, once a division cycle begins, cells are committed to completing it.

The genome of *S. cerevisiae* consists of 16 small chromosomes, ranging in size from 350 to 1400 kbp, with a total haploid DNA content of about 1.5×10^4 kbp. The yeast mitochondrion, which carries out the enzyme reactions in aerobic respiration, has a separate circular chromosome of about 75–80 kbp, of which 10–40 copies may be present, depending on the physiological state of the cell [33,34]. For comparison, the DNA content of a typical bacterium, such as *E. coli*, is about 0.5×10^4 kbp, all in a single circular chromosome, while the human genome contains approximately 3×10^6 kbp [32]. In comparison with most other higher organisms, then, yeast chromosomes are extremely small and may be visualized intact after electrophoretic separation in agarose gels.

Although the nuclear chromosomes in yeast are precisely maintained, duplicated, and assorted during cell division, the mitochondrial chromosome is quite unstable. In about 1–3% of divisions, the daughter inherits a defective chromosome, carrying duplications and deletions of portions of the chromosome. Such cells are consequently unable to respire, although they continue to live and divide and can carry out fermentation.

Many yeasts have the potential to reproduce sexually. Under conditions of nitrogen starvation and the absence of fermentable carbon, cells of *Saccharomyces* may enter the sexual phase and undergo meiosis, as long as they contain two or more complete sets of chromosomes. In this case, cells give rise to one to four (sometimes eight) spores with a reduced chromosome complement: diploids, with two sets of chromosomes, produce haploid spores with one set each, whereas tetraploids, with four sets, produce diploid spores. Commercial yeast strains are at least diploid. Ability to sporulate is determined by the mating type locus: cells carrying only the *MATa* or *MATα* allele are able to mate with cells carrying the other allele but not to sporulate; cells carrying both alleles (MATaα) cannot mate but may sporulate. However, after repeated passage through the vegetative state, as happens with sourdough yeast starters, most yeasts eventually lose the ability to sporulate no matter what their mating type.

The yeast cell wall, which makes up about 30% of the total cell solids, consists of a matrix of protein-linked mannose and glucose chains, and two glucose-containing molecules, a β1,3-glucan and a β1,3-glucan [51]. The outer wall consists mainly of mannans, which are mannose-containing proteins with extensive O- and N-linked carbohydrate chains, and the inner wall mainly of fibrous, helical β1,3-glucan. The two are linked by the β1,3-glucan. Some chitin, about 120 sugar residues long, is also found [52]. Certain carbohydrate-linked molecules, such as invertase and phosphatase, are secreted by the cell and remain in the periplasmic space between the cell wall and the cell membrane. There they act on substrates that diffuse through the cell wall and modify them for uptake by the cell. The cell wall makes the cell extremely resistant to mechanical damage and prevents the cell from swelling in hypo-osmotic environments.

III. YEAST METABOLISM

A. Carbohydrates

As living organisms, yeasts in doughs and ferments take up nutrients, especially carbohydrates (sugars) as sources for energy and growth (see Ref. 3). Most important from the baker's point of view is the fermentative metabolism of carbohydrates, which provides the carbon dioxide necessary to leaven dough, along with ethanol (ethyl alcohol), which plays a role in conditioning flour proteins.

Flour contains a large number of carbohydrates [41,42]. In addition to the major starch constituents amylose and amylopectin, wheat contains about 5% pentosans (sometimes referred to as hemicelluloses), 1–2% glucofructans, and 1–2% mono-, di-, and trisaccharides. Amylose and amylopectin, both polymers of the six-carbon sugar glucose, are found in the starch granules. Amylose is a linear polymer of glucose molecules; in amylopectin this polymer is complicated by the addition of glucose side chains. Starch granules that have been damaged during milling are accessible to the enzymes α- and β-amylase, which degrade both amylose and amylopectin to produce glucose and maltose, a disaccharide containing two glucose molecules.

The pentosans are also polymers, containing the five-carbon sugars (pentoses) xylose and arabinose, as well as galactose and small amounts of glucose. Glucofructans are chains of fructose molecules appended to a single molecule of glucose. Sucrose may be regarded as the smallest possible glucofructan. The smaller carbohydrates include sucrose (up to 1% by weight of the flour) and raffinose, as well as small amounts of glucose and fructose.

Baker's yeasts in general are capable of utilizing a variety of sugars present in flour: glucose, fructose, mannose, maltose, and sucrose. On the other hand, they do not utilize the pentoses xylose or arabinose, and consume raffinose only partially. In addition, *S. exiguus* is unable to take up maltose, a trait it may have developed during the development of its symbiotic relationship with the sourdough lactobacterium *L. sanfrancisco*.

Yeast cells possess a variety of transport enzymes to bring sugars from the environment into the cell where they can be metabolized. The monosaccharides glucose and fructose, molecules consisting of a single sugar unit, can be used directly, whereas the disaccharides sucrose (glucose plus fructose) and maltose (glucose plus glucose) must be broken down to their component monosaccharides before they are metabolized. Glucose-sensing molecules (receptors) in the cell membrane monitor the presence of glucose and signal the cell to synthesize hexose transporters and begin importing glucose for energy production [36]. The sensor molecules themselves, though closely related to transporters, are specialized for signaling and do not transport glucose themselves [37]. Glucose and maltose are naturally present in flour, and additional amounts are released by the action of flour and bacterial amylases on broken or damaged starch granules. Sucrose or fructose may be added by the baker as table sugar or syrup.

Glucose and fructose are the first sugars to be used during fermentation. Sucrose is rapidly broken down to glucose and fructose by yeast enzymes already present outside the cell membrane, whereas maltose is not consumed until glucose is almost exhausted. Because amylase, a starch-degrading enzyme in flour, is constantly generating new glucose and maltose from flour starch, maltose levels may actually increase in the early stages of fermentation in a yeast-only dough.

The accumulation of maltose results from one of the most important *regulatory* mechanisms in yeast: *glucose repression* (see Ref. 26 for a recent review). The presence of glucose, fructose, or mannose in dough at concentrations above 0.1% leads to a long series of adaptations by the yeast cells. First, they generate energy by fermentation rather than respiration. Although this is inefficient from the standpoint of generating energy and producing new yeast cells, it results in very rapid production of carbon dioxide and ethanol. Second, enzymes and structures such as mitochondria that are necessary for respiration are degraded or reduced to rudiments. Cells become unable either to respire glucose or to use such nutrients as glycerol, ethanol, or acetate for energy. Third, yeast cells stop manufacturing internal storage carbohydrates (such as trehalose and glycogen), and rapidly break down and metabolize any that may be present. Fourth, enzymes required for the uptake and utilization of other sugars in the dough, especially maltose, are degraded and their synthesis terminated (since glucose and fructose are taken up by the same enzyme [4], fructose uptake is not affected). Fifth, glucose activates a number of enzyme systems used to transport nutrients to the cell interior, including amino acids [5] and ammonium [6].

Negative control is mediated by a wide variety of mechanisms, including transcriptional repression, mRNA degradation, decreased translational efficiency, protein degradation, enzyme inactivation, and enzyme inhibition. The initial regulatory step is carried out by the enzyme hexokinase, whose principal function—addition of a phosphate group to glucose—is the first step in sugar metabolism. In the presence of glucose, hexokinase can also add phosphate to other enzymes, and in this way it initiates a regulatory cascade by activating a second enzyme that in turn inhibits a third enzyme (SNF1 protein) whose function is to prevent synthesis of MIG1 protein, the major transcriptional repressor of glucose-repressible genes. Among the metabolic pathways repressed by MIG1 are those involved in galactose, maltose, and sucrose uptake and metabolism, gluconeogenesis, and the respiratory chain. It is important to know that some or all enzymes are regulated in more than one way by glucose. For example, transcription of the trehalose synthetase genes is repressed by MIG1, and the enzymes themselves are inactivated by addition of phosphate, a process also controlled by hexokinase [27]. Likewise, the genes required for uptake and metabolism of maltose are turned off by MIG1, and the proteins present when glucose appears are marked for destruction by a second set of enzymes induced in the presence of glucose [35]. The purpose of these complex control mechanisms appears to be amplification: because each enzyme in the regulatory cascade can act on more than one enzyme in the subsequent step, the presence or absence of only a few molecules of glucose in the cell leads to large and rapid changes in the cell's overall metabolic state.

The end result of this process is that molecules of glucose and fructose are taken up by the yeast cells and transformed by *glycolysis*. Normally, each sugar molecule produces two molecules of carbon dioxide and two of ethanol, and the yeast cells obtain enough energy to maintain their metabolism as well as very limited growth. Working out the steps of glycolysis was the first major accomplishment of the science of biochemistry and was based on observations made by Harden and Young on crude yeast extracts ("zymase"). It was found that each six-carbon glucose or fructose molecule is split into two three-carbon sugars (trioses), which are subsequently rearranged, oxidized, and cleaved into the two-carbon alcohol ethanol and the one-carbon molecule carbon dioxide. These rearrangement and cleavage reactions generate energy for the yeast. The enzymes responsible for these reactions may make up 50% of the soluble protein in actively fermenting cells [7]; in the early stages of fermentation, the rate of sugar consumption is limited only by the cell's ability to transport sugars across the cell membrane [8]. Since the yeast cell wall makes up 20–30% of the total cell solids, it is clear that actively fermenting yeasts devote an enormous share of their activity to glycolysis.

In addition to ethanol and carbon dioxide (the two major products of glycolysis), glycerol, acetic acid, and succinic acid are always produced at low levels during fermentation: they may account for 3–5% of the sugar consumed, in a roughly constant molar ratio in which glycerol formed is equal to twice the acetate plus five times the succinate [8]. It is generally accepted that the production of these compounds is required for growing cells to maintain their internal redox balance and provide the reducing equivalents necessary to support synthesis of new cell constituents.

In a growing culture, carbohydrates are required for many uses besides energy production. The yeast cell wall consists of a complex array of long-chain glucose-containing carbohydrates and includes a great many glycoproteins as well, mostly containing mannose [30]. Excreted and membrane-bound proteins are all modified by the addition of carbohydrate side chains, mainly polymers of mannose. In the absence of an external source of amino acids, carbon skeletons derived from glucose are required for protein synthesis. Finally, nucleic acids, which make up 8–12% of the cell weight, all contain five-carbon sugars that must be synthesized from a glucose precursor.

B. Nitrogen

Regarding the role of nitrogen in yeast metabolism, the reader is referred to Refs. 28 and 29 for reviews. In addition to utilizing sugars for energy and protein synthesis, yeasts require a source of amino nitrogen, most of which is devoted to protein synthesis. Nitrogen uptake is performed by a variety of membrane-bound enzymes (permeases) with specificities and affinities for ammonia, different amino acids, amino acid families, and short oligopeptides. In the presence of ammonia, glutamic acid, or glutamine, nonspecific amino acid permeases are repressed until intracellular glutamine levels rise and the cell disables the repression. At this point, the cell regains the ability to take up any free amino acids from its environment. This change, controlled by the *URE2* gene, allows yeasts to grow rapidly on media containing a rich source of amino acids. In addition, the same regulatory network represses enzymes used to synthesize and interconvert most amino acids as long as they can be obtained externally [29]. In a manner reminiscent of carbohydrate uptake, the presence of amino acids appears to be sensed by specialized membrane proteins, structurally related to the membrane proteins that actually transport amino acids into the cell [38].

Flour contains a significant amount of free amino acids, and the action of flour and bacterial proteases in an active sourdough generates sufficient peptides to support the growth of the cells. Lactobacterial proteases are extremely active and produce much more amino nitrogen than the bacteria consume themselves [9]. In addition, the yeasts will themselves excrete some amino acids, especially during early growth [25]. In general, only glutamic acid and glutamine continue to be synthesized at high levels because of their importance in regulating the intracellular redox balance.

C. Vitamins

Most yeasts must be supplied with biotin. In commercially grown yeasts this is supplied by cane molasses in the growth medium. Wheat flour contains sufficient biotin to support yeast multiplication and fermentation. Addition of pyridoxine, inositol, and pantothenic acid may increase growth rates, and thiamine is added to commercially grown cells to increase their fermentation rate. In general, however, yeasts are an excellent dietary source of B vitamins and are capable of meeting their own needs by synthesizing vitamins from simple precursors.

IV. FLAVOR COMPOUNDS

The great appeal of sourdough breads lies in their characteristic sour flavor. Although the major flavor is contributed by the lactobacterial acids, primarily acetic and lactic, this flavor is modulated by contributions from a number of other flavored compounds. As shown in Table 2, a large number of biogenic flavor chemicals are elaborated by yeast in sourdough; many of these are also synthesized by the lactobacteria. The amounts produced depend significantly on which yeast and lactobacterium are present in the sourdough: differences of up to sixfold in the concentrations of such compounds have been measured in experimental doughs. Several compounds are produced only in mixed cultures of yeast and bacteria, and it is commonly observed that the yeast produces larger amounts of many compounds, such as propionic acid, in the presence of the bacteria than in their absence. The table also shows that the minor flavor compounds contribute flavors in addition to the primary sour taste; fruity, buttery, and other flavor notes are found and do contribute to the overall acceptability and attractiveness of the product. Some of these are present at levels below their detection threshold and are sensed only in combination with related compounds.

Table 2 Yeast-Produced Flavor Compounds in Sourdoughs

Compound	Flavor	Reported conc. (mg/kg)	Perception threshold (mg/kg)
Alcohols			
Ethanol[a]	Alcoholic		
n-Propanol[a]	Fusel-like, burning		
2-Methyl-1-propanol[a]			
2-Butanol[b]			75.0
2-Methyl-1-butanol			
3-Methyl-1-butanol			
n-Pentanol (amyl alcohol)	Fusel-like, burning		
i-Pentanol (*i*-amyl alcohol)[a]			6.5
n-Hexanol[a]	Alcoholic		
Carbonyls			
Acetaldehyde	Pungent		1.2
Propionic acid[a]	Rancid	6–127 (w/bacteria)	
n-Butyric acid[a]	Rancid butter		
i-Butyric acid[a]	Sweaty	20	
n-Valeric acid[a]	Rancid butter		
i-Valeric acid[a]		6–8	0.75
Hexanoic acid	Unpleasant, copra-like	2–4	
Acetone[a]	Aromatic, sweet		
Methylpropanal	Malty	1.7–4.7	
2-Methyl-1-butanal	Malty	1.1 1.1	
3-Methyl butanal	Malty	0.4–0.8	
2,3-Butanedione (diacetyl)	Butter		0.02
3-Hydroxy-2-butanone (acetoin)[b]	Butter	1.5	
n-Hexanal[a]	Fruity		
trans-2-Heptenal[a]	Green, fatty		
Methional	Malty	0.03–0.04	
Esters			
Ethyl acetate[a]	Ether, pineapple		17.0
2-Acetyl-1-pyrroline	Roasty	0.0119–0.0159	
i-Amyl acetate	Fruity		
Phenethyl acetate	Fruity		
2,3-Methylbutyl acetate	Apple peel, banana		
n-Pentyl acetate[b]			
n-Hexyl acetate[b]	Pear, bittersweet		0.67
Ethyl *n*-propanoate[b]	Rum, pineapple		
Ethyl *n*-hexanoate[a]	Pineapple, banana		0.008–0.076
Ethyl lactate[a]			14
Alkanes			
Heptane			
Octane			

[a] Also produced by sourdough lactobacteria.
[b] Detected only in mixed lactobacterial/yeast fermentations.
Source: Refs. 16, 19, 20, 22, 54, and 58.

More than 100 flavor chemicals have been identified in bread, most of them present in minor amounts (see Ref. [60] for a typical analysis). These include higher alcohols, aldehydes, acids, alkanes, and esters. The alcohols ("fusel oils") are produced by deamination of amino acids and are produced most rapidly when the yeasts are starved for a source of amino nitrogen. Their total concentration in dough is less than 100 ppm, and their individual perception thresholds are about the same. Acetaldehyde, the most common aldehyde, is an intermediate in glycolysis; most is transformed enzymatically into ethanol but some escapes from the cell. The ester ethyl acetate results from a reaction between ethanol and coenzyme A inside the yeast cell.

The other minor flavor compounds are produced as byproducts of yeast secondary metabolism, especially involving the interconversion of amino acids. 3-Methyl butanal, which produces a malty flavor, is produced from leucine [59], and suggestions have been made that increasing the level of leucine in dough by direct supplementation would increase the malty flavor of the resulting bread [58]. These secondary metabolites may be further transformed by the cell. Such changes depend sensitively on the particular strain of yeast in use. Some interest has appeared recently in adding small amounts of wine and brewer's yeast to dough fermentations to enhance the flavor profile. A number of mutant yeast strains with altered flavor profiles have been developed by the wine making industry [53] by selecting for resistance to amino acid analogs. In addition to alterations in the levels of fusel oils, these strains show changes in acetaldehyde and acetic acid as well.

Factors such as yeast level, fermentation time, and fermentation temperature all affect the levels of flavor compounds in bread. Not only does an increase in fermentation time allow the yeast to produce more metabolites; it also alters the balance of flavors because as fermentation time increases dough proteases generate more free amino acids to react with the yeast metabolites. Thus, an increase in the malty flavor contributed by methyl propanal, 2-methyl butanal, 3-methyl butanal, and methional as fermentation time increased was attributed to an increase in free amino acid availability, whereas the fatty flavor contributed by 1-octen-3-one and (*E*)-2-nonenal, derived from flour fatty acids, did not show any change [58]. In the same study, an increase in fermentation temperature led to a decrease in the level of the malty flavored compounds, with no effect on the fatty flavors, and an increase in the yeast level led to an increase in the roasty flavored 2-acetyl-1-pyrroline and only minor changes in the other flavor compounds.

The final flavor is produced in a two-step process. Primary components are synthesized during the fermentation. However because the last stage in breadmaking is baking, many of the fermentation products, especially the volatiles, evaporate or are further modified by the high temperature in the oven. The volatiles are not all lost from the loaf, however. Because the dough is heated from the outside in, volatiles tend to be concentrated in the interior of the loaf as baking progresses. For example, a freshly baked bread may still contain up to 0.5% (flour weight) ethanol, despite the fact that its internal temperature is well above the boiling point of ethanol.

The most important of the flavor-producing modifications occur as a result of Maillard reactions between sugar hydroxyl groups on one molecule and amino nitrogen on a second (typically an amino acid or short peptide derived from flour protein). For example, the flavor compound 2-acetylpyrroline is formed by the reaction of the amino acid ornithine with the yeast metabolite pyruvaldehyde [59]. Many of these reactions occur during the browning of the crust, and the newly synthesized flavor molecules slowly diffuse out of the crust into the body of the loaf. Removing the crust of a newly baked bread results in the loss of the expected fermented flavor. Generally, concentrations of flavor compounds are higher in the crust than in the body of the loaf [60]. Temperatures in the crust may reach 150–180°C, whereas the interior, because of the high water content, remains at 99–100°C. For this reason, a typical baked flavor is absent in microwaved or steamed breads [22].

V. YEAST ACTION ON PROTEINS

The major flour proteins are the glutenins and the gliadins (see Ref. 61 for a recent review). The glutenins are a family of related molecules divided into high and low molecular weight forms. Both forms contain a large glutamine-rich central region containing a repeating amino sequence with shorter (50–100 amino acids) cysteine-containing domains at each end. The high molecular weight forms are further subdivided into x and y types. In flour, the x and y molecules are always found linked head-to-tail by disulfide bonds, never as single monomers. The number of such disulfide bridges that any glutenin can form depends on the number and arrangement of cysteine residues at the ends of the molecule. Generally, the carboxy terminus contains a single cysteine and can form only one disulfide bond; the amino terminus contains a variable number (up to 5) and may thus participate in more than one, creating branch points in the overall complex. There is a general tendency for cultivars of wheat with glutenins containing multiple cysteines to produce flour with stronger mixing characteristics [62]. Presumably this means that strength is contributed by highly branched glutenin complexes rather than by long, linear forms. However, linear forms may produce a more extensible dough. In any case, this suggestion lends support to the idea that rearrangement of intermolecular disulfide bonds is an important factor in dough develpment.

The overall structure of the low molecular weight glutenins resembles that of the high molecular weight forms but contains a different internal repeating sequence. The low molecular weight forms do not contain enough cysteine residues in their terminal regions to form more than one intermolecular disulfide bond and thus can participate only in the creation of linear chains of gluten subunits. Not surprisingly, wheat cultivars such as durum, with a high proportion of low molecular weight gluutenins, are not good bread wheats but make flour with good extensibility.

High and low molecular weight glutenin molecules may also differ in the length of their internal repeat-containing region. Some experiments indicate that glutenins with longer repeat regions produce stronger and more extensible doughs [61].

Glutenins and gliadins both contain a high proportion of the basic amino acid glutamine, which appears to play an important role in maintaining the native structure of the proteins through the formation of hydrogen bonds between (and within) molecules. The decrease in pH due to acid production during fermentation must assist in disrupting this structure and denaturing the proteins. In addition, the alcohol produced by fermentation is itself a powerful denaturing agent, especially in an acid environment. Approximately one-half by weight of the sugar consumed by yeast during fermentation is released as alcohol: in a lean dough with no added sugar, this may amount to 0.5–1.0% of the dough, and as much as 3–5% in a dough with 5–10% added sugar. The presence of this much alcohol may affect the colloidal state of the flour proteins. The maturation of a dough appears to require significant changes in the conformation of the dough proteins, e.g., from a mainly helical or coiled conformation in the flour to an elongated, linear shape in the final dough. The combined effects of pH and alcohol in inducing this change and then in preventing the proteins from regaining their original shape are clearly important. Yeast cells do not secrete any enzymes, such as proteases, that might directly affect dough protein structure. These enzymes are present in the flour and are also produced and secreted by bacteria in the starter. Lactobacteria outnumber yeast at all stages of a soda cracker fermentation [48] by about 3:1; total microbial counts are about 7×10^8 per gram.

Soda cracker production resembles sourdough production: an overnight fermentation is used, inoculated from a starter that is renewed by addition of flour and water. Dough reconstitution experiments showed that changes in dough rheology during the early stages of fermentation in a mixed yeast/lactobacterial soda cracker dough involve alterations in the insoluble gluten fraction [43]. The lack of change in the viscosity of this fraction suggested that the proteins had not been degraded and led to the conclusion that it was the protein conformation that had changed. While these rheological changes could be reproduced at high pH by the addition of glutathione and other reducing agents, such chemical effects disappeared at the low-pH characteristic of sourdoughs. Thus, the early changes in protein structure may be due to factors other than proteolysis or the reduction of disulfide bonds.

Alterations in cracker dough rheology during the later stages of fermentation appear to depend on the action of proteases. When flour is treated with a solution of ammonium sulfate to remove soluble enzymes, the remaining solids, when incorporated into a dough, do not display the normal increase in elasticity [44]. A dough model system made by mixing purified gluten with purified starch exhibits the same stability [45]. Addition of the crude ammonium sulfate enzyme extract or of the substantially purified protease to either of these dough systems produces the normal extensibility changes. Furthermore, in both dough systems the effect of the enzyme extract depends on dough pH, with the maximal effect occurring at pH 3.8–4.1. This pH is reached in a typical cracker dough fermentation about 9 h after mixing. A typical cracker dough fermentation lasts about 18 h, or 9 additional hours after this point is reached. A mock cracker dough system using no starter with a 3-h yeast fermentation followed by addition of lactic acid to bring the pH to 3.8–4.0 and a subsequent 12-h incubation produced crackers with substantially normal physical qualities [46]. A similar effect was produced by inoculating the cracker sponge with a lactobacterial starter culture to give an initial lactobacterial level of about 10^8 cells per gram [49,50]. This emphasizes the importance of pH alone in developing a dough with the proper rheology.

The carbon dioxide produced by fermentation dissolves into the dough in the form of bicarbonate as it is excreted by the yeast and acidifies the dough as well. Accumulation of carbon dioxide in bubbles does not begin until the dough is saturated with bicarbonate. Bakers will occasionally add ammonium salts, such as ammonium sulfate or ammonium chloride, to doughs in order to increase fermentation rates. Assimilation of the ammonium in these salts by the yeast leads directly to the generation of sulfuric or hydrochloric acid. In addition, the acids excreted by yeast, though not as plentiful as the bacterial acids in a sourdough, will affect the pH of the dough and probably have some effects on protein structure as well as promoting protease activity.

Carbon dioxide produced by the yeast during fermentation collects within the gas mainly in the form of very small, finely divided bubbles. At the surface of a growing bubble, proteins are subjected to very strong stretching forces.

VI. FERMENTATION RATES AND LEAVENING ACTION

During a standard bakery fermentation, the sugar consumed by yeast amounts to roughly 3% of the flour weight, whether the dough is supplemented with added sugar or not [55,56]. (Sugar beyond this level is not consumed and ends up in the final loaf.) For a standard 1-pound loaf, this amounts to almost 5 g of carbon dioxide, or a volume of about 2500 cm^3 (over 1/2 gallon) of carbon dioxide gas. An equivalent weight of ethanol is produced at the same time. Most of the carbon dioxide (80–100%) and almost all of the ethanol is retained in the dough until the final baking, when the loaf undergoes an expansion of about one-third to a final volume of about 2300 cm^3. About one-half of this increase is contributed by carbon dioxide: of this, 57% results from expansion of preexisting gas in bubbles and 39% by carbon dioxide forced out of solution by the increase in temperature. The rest of the expansion is accounted for by the evaporation of ethanol [57].

The main factors affecting fermentation rates are temperature and osmotic pressure. Figure 1 shows yeast fermentation rates for three doughs at different sugar concentrations. Initially, the lean dough (no added sugar) shows the most rapid CO_2 production; however, as the preexisting glucose in the dough is rapidly exhausted, the fermentation rate drops off after

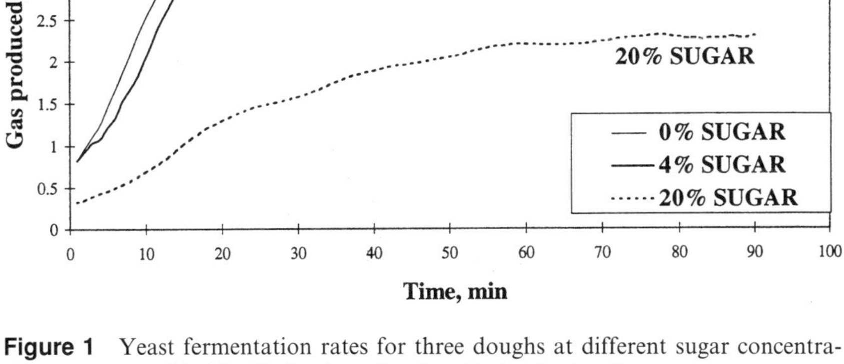

Figure 1 Yeast fermentation rates for three doughs at different sugar concentrations. Data provided by E. Sanders, Sensient Technologies Corp.

about 15 minutes. Glucose liberated from starch by amylases continues to support leavening at a reduced level until about 50 minutes into the fermentation, when the yeasts finally begins to utilize maltose, leading to a gradual increase in CO_2 production. *Saccharomyces exiguus*, which does not ferment maltose, would not produce this second peak of activity. Doughs containing 4% added sucrose initially ferment at a slightly lower rate; however, since the sugar is not exhausted, fermentation continues without any decrease in activity. A significant inhibition of fermentation is seen in the sweet dough (20% added sugar), which begins fermentation at a much lower rate. During the course of the fermentation, a continual increase in the rate of CO_2 evolution occurs since the osmotic pressure of the dough gradually drops off as sugar is consumed. A similar phenomenon is observed when salt is added to dough: salt concentrations above 1.5% depress the rate of fermentation (although salt additions of 0.5–1% may increase the fermentation rate).

For commercial yeasts, the optimal fermentation temperature is about 38°C, although the best temperature varies with different dough formulations. The sensitivity of yeasts to temperature depends a great deal on the nature of the dough. For example, sweet doughs (i.e., 15–20% added sugar) are much more sensitive to high temperatures because the high osmotic pressure produced by the sugar already subjects the yeasts to significant stress. In addition to changes in the rate of fermentation, which is generally higher at higher temperatures, significant differences may be observed in flavor profiles.

In the pH range of 4–6, only slight changes in fermentation rate occur, although fermentation drops off quickly below pH 4. Bread doughs are very well buffered by the flour proteins, and the pH rarely falls below 4 except in some extremely sour ferments, when it may drop to 3.8 [31]. The change in pH is brought about by the excretion of organic acids by the yeast and the lactobacteria. Yeast secretes acetic and succinic acids during fermentation; in addition, some of the carbon dioxide produced by fermentation dissolves to form bicarbonate ions, which also acidify the dough.

Maximum fermentation rates for commercial baker's yeasts range from 0.3 to 0.7 g carbohydrate consumed per gram yeast solids per hour [31]. This rate is reached only gradually. Yeast fresh from the manufacturer is in a resting state and requires time to revive and adjust to its new environment. The time required depends primarily on fermentation temperature and ranges from 30 min at 107°F to 150 min at 84°F [31]. Thus, a baker's yeast added to a dough will not reach maximal activity until well after the dough is mixed.

In addition to the factors discussed above, yeasts themselves possess quite variable abilities to ferment different sugars. So-called lean dough yeasts, typically used in doughs with little or no added sugar, display relatively high basal levels of maltase activity, even in the absence of maltose, and sometimes even in the presence of glucose at concentrations of 8% or more. However, such strains typically perform less well in doughs supplemented with sugar at this level and display very poor activity in sweet doughs due to inhibition by high osmotic pressure. Yeasts chosen to perform well in sweet doughs generally perform less well that lean dough yeasts in dough with 8% added sugar and very poorly in lean doughs.

The accumulation of carbon dioxide in the dough after mixing occurs in two phases: as dissolved bicarbonate, and as gaseous CO_2 in bubbles. The rising of the dough during proofing results from the growth of these bubbles; carbon dioxide is only slightly soluble, and as more is produced by the fermenting yeast, bicarbonate is forced out of solution. Some of the gas may escape through the surface of the dough, but the majority is trapped in the body of the dough piece and is forced into preexisting bubbles. It is the growth of these bubbles that produces the increased volume of a proofed dough. Gas retention depends on the development of the proper dough structure, which requires adequate dough mixing.

A final and dramatic increase in dough volume occurs when the dough is placed in the oven. This "oven spring" results from a further increase in bubble volume as the trapped gas expands due to heating; from the decreased solubility of carbon dioxide at increased temperatures; from the evaporation of alcohol in the dough; and from the increased activity of yeast enzymes at higher temperatures before they are finally inactivated at about 40–50°C.

REFERENCES

1. P. E. McGovern, D. L. Glusker, L. J. Exner, and M. M. Voight, *Nature*, 301:480 (1996).
2. D. Samuel, *Science*, 273:488 (1996).
3. C. Gancedo and R. Serrano, *The Yeasts, 2nd ed.* (A. E. Rose and J. H. Harrison, eds.). Academic Press, New York, 1989, p. 205.
4. C. F. Heredia, A. Sols and G. De la Fuente, *Biochim. Biophys. Acta*, 86:216 (1968).
5. E. H. Ramos, L. C. DeBongioanni, M. C. Cuesta-Casedo and A. O. M. Stoppani, *Biochim. Biophys. Acta*, 467:220 (1977).
6. R. J. Roon, H. L. Evan, P. Dunlop and F. L. Larimore, *J. Bacteriol.*, 122:502 (1975).
7. B. Hess, A. Boileux and J. Kruger, *Advances in Enzyme Regulation* (B. Wright, ed.). Ronald Press, New York, 1969, p. 149.
8. E. Oura, *Process Biotechnology*, 12:19 (1977).

9. C. Collar, *Food Sci. Technol. Int.*, 2:349 (1966).
10. M. Rohrlich, *Baker's Digest*, 35:44 (1966).
11. A. Schulz, *Baker's Digest*, 40:77 (1966).
12. S. Barber, M. J. Torner, M. A. Martinez-Anaya, and C. B. de Barber, *Z. Lebensm. Unters. Forsch.* 189:6 (1989).
13. T. F. Sugihara, L. Kline, and M. W. Miller, *Appl Microbiol.*, 21:456 (1971).
14. G. Spicher, E. Rabe, R. Sommer, and H. Stephan, *Z. Lebensm. Unters. Forsch.*, 173:291 (1981).
15. P. Damiami, M. Gobetti, L. Cossagni, M. S. Simonetti, and J. Rossi, *Lebensm. Wiss. Technol.*, 29:63 (1996).
16. M. A. Martinez-Anaya, B. Pitarch, P. Bayarri, and C. B. de Barber, *Cereal Chem.*, 67:85 (1990).
17. J. A. Barrett, R. W. Payne, and D. Yarrow, *The Yeasts: Characteristics and Identification.* Cambridge University Press, Cambridge, 1983, p. 417.
18. D. Yarrow, *The Yeasts: A Taxonomic Study* (N. J. W. Kriger-van Rij, ed.). Elsevier Science, Amsterdam, 1984, p. 379.
19. B. Hansen and A. Hansen, *Z. Lebensm. Unters. Forsch*, 198:202 (1994).
20. M. A. Martinez-Anaya, M. J. Torner, and C. B. de Barber, *Z. Lebensm. Unters. Forsch*, 190:126 (1990).
21. G. A. Burdick, *Fenaroli's Handbook of Flavor Ingredients, 3rd ed.* CRC Press, Boca Raton, 1994.
22. G. R. Reed and T. W. Nagodawithana, *Yeast Technology, 2nd. ed.* Van Nostrand Reinhold, New York, 1991.
23. T. F. Sugihara, in G. J. B. Wood, ed., *Microbiology of Fermented Foods, Vol. 1.* Elsevier, New York, 1985.
24. Y. Oda, M. Yabuki, K. Tonomura and M. Fukunaga, *Yeast*, 13:1243 (1977).
25. K. Lorenz, *Bakers Digest*, 60:41 (1983).
26. C. J. L. Klein, L. Olsson and J. Nielsen, *Microbiology* 144:13 (1998).
27. M. A. Blazquez and C. Gancedo, *Curr. Genet.* 25:89 (1994).
28. E. W. Jones and G. R. Fink, *The Molecular Biology of the Yeast Saccharomyces: Metabolism and Gene Expression.* Cold Spring Harbor Press, New York, 1982, p. 181.
29. B. Magasanik, *The Molecular Biology of the Yeast Saccharomyces: Gene Expression, 2nd Ed.* Cold Spring Harbor Press, New York, 1992, p. 283.
30. C. E. Ballou, *The Molecular Biology of the Yeast Saccharomyces: Metabolism and Gene Expression.* Cold Spring Harbor Press, New York, 1982, p. 335.
31. E. J. Pyler, *Baking Science and Technology, 2 Vols., 3rd ed.* Sosland Publishing, Merriam, KA, 1988, p. 246.
32. B. Lewin, *Gene Expression 2: Eukaryotic Chromosomes, 2nd ed.* John Wiley & Sons, New York, 1980, p. 490.
33. B. Stevens, *The Molecular Biology of the Yeast Saccharomyces: Life Cycle and Inheritance.* Cold Spring Harbor Press, New York, 1982, p. 471.
34. T. D. Fox, L. S. Folley, J. J. Mulero, T. W. McMullin, P. E. Thorsness, L. O. Hedin and M. C. Costanzo, *Methods in Enzymology, Vol. 194: Guide to Yeast Genetics and Molecular Biology*, Academic Press, San Diego, 1991, p. 149.

35. T. Harma, et al., *J. Biol. Chem.* 273:15352 (1998).
36. P. Vagnoli, D. M. Coons and L. Bisson, *FEMS Microbiol. Lett.* 160:31 (1988).
37. S. Ozcan, J. Dover and M. Johnston, *EMBO J.* 17:2566 (1988).
38. J. Winderickx et al., *Molec. Gen. Genetics* 252:470 (1996).
39. V. Mantynen et al., *System. Appl. Microbiol.* 22:87 (1999).
40. M. Almeida and C. Pais, *Appl. Env. Microbiol.* 62:4401 (1996).
41. D. W. Kent-Jones and A. J. Amos, *Modern Cereal Chemistry, 6th ed.* Food Trade Press, London, 1967.
42. R. C. Hoseney, *Principles of Cereal Science and Technology.* American Association of Cereal Chemists, St. Paul, MN, 1986.
43. J. Y. Wu and R. C. Hoseney, *J. Food Proc. Preserv.* 14:17 (1990).
44. W. D. A. Lin, G. Lookhart and R. C. Hoseney, *Cereal Chem.* 70:448 (1993).
45. J. Y. Wu and R. C. Hoseney, *Cereal Chem.* 66:182 (1989).
46. D. E. Rogers and R. C. Hoseney, *Cereal Chem.* 66:6 (1989).
47. C. P. Kurtzman and J. W. Fell, eds., *The Yeasts: A Taxonomic Study.* Elsevier, New York, 1998.
48. T. F. Sugihara, *J. Food Prot.*, 41:977 (1978).
49. T. F. Sugihara, *J. Food Prot.*, 41:980 (1978).
50. T. F. Sugihara, *Soda Crackers.* U.S. Dept. of Agriculture, National Tech. Information Service, PB80-141427, Washington, DC, 1979.
51. P. N. Lipke and R. Ovale, *J. Bacteriol.* 180:3735 (1998).
52. R. E. Kollar et al., *J. Biol. Chem.* 270:1170 (1995).
53. G. Suzzi, P. Romano and M. Polsinelli, *World J. Microbiol. Biotech.* 14:243 (1998).
54. D. Maloney, *Kirk-Otthmer Encyclopedia of Chemical Technology, 4th ed.*, Vol. 26, 1998, p. 761.
55. S. S. Jackel, *Proc Am. Soc. Bakery Engrs.*, 91 (1969).
56. E. R. Peikarz, *Proc Am. Soc. Bakery Engrs.*, 118, (1963).
57. W. R. Moore and C. R. Hoseney, *Cereal Foods World*, 30:791 (1985).
58. G. Zehenbauer and W. Grosch, *J. Cereal Sci.*, 28:93 (1996).
59. I. Labuda, C. Stegmann and R. Huang, *Cereal Foods World*, 42:797 (1997).
60. L. M. Steitz, O. K. Chung and R. Rengarajan, *Cereal Chem.*, 75:847 (1998).
61. D. D. Kasarda, *Cereal Foods World*, 44:567 (1999).
62. P. I. Payne, M. A. Nightingale, A. F. Krattiger and L. M. Holt, *J. Food Sci. Agric.* 40:51 (1987).

4

Associations and Interactions of Microorganisms in Dough Fermentations: Effects on Dough and Bread Characteristics

M. Antonia Martínez-Anaya
Instituto de Agroquímica y Tecnología de Alimentos (CSIC), Valencia, Spain

I. INTRODUCTION

Sourdough is a characteristic example of association of lactic acid bacteria and yeasts. Its functionality lies in a suitable balance between lactic and alcoholic fermentation. Sourdough processes are a common practice in the preparation of traditional rye [1,2] and wheat flour breads [3–21]. In rye bread production they are necessary to obtain technical quality, development of bread flavor, inhibition of undesirable fermentations, activation of yeast, and inhibition of mold and rope growth. In wheat bread production, sour doughs mainly help to develop bread flavor and to delay bread spoilage [22].

The greatest difficulty in studying complex biological systems such as sourdoughs results from the numerous interactions between the environment, microorganisms, and technological treatments [23,24]. Sourdough processes are affected by many internal and external factors. Internal factors include the composition of the microflora (type, number, growth requirements, fermentation and enzymatic activities, etc.) [1,16,22,25–27], and flour characteristics (extraction rate, nutrients, minerals, growth activators, enzymatic activity, etc.) [13,22,27–30]. External factors refer to dough consistency, temperature, fermentation time, oxygenation, addition of yeast, ingredients, and others [13,16,26,28–32].

A sourdough process can be started (1) spontaneously, resting a mixture of flour and water under suitable conditions; (2) using a well-defined microbial starter; or (3) adding a seed sour from a previous batch [1,14,15,22]. Full sourdoughs to be used in bread dough mixing can be obtained after a single or a multistage process [22,33]. The sour dough acts as an improver in rye and wheat flour systems. Reported effects include to shorten fermentation times, reduce fermentation losses, speeds dough ripening, increase gassing power, improve machinability of doughs, and raise acidity of doughs and breads [2,34–38]. Also, loaves have better volume and sensory scores [1,3,4,6,9,13,29,39–42], and physical and microbiological shelf life is extended [13,28,29]. If used, the quality of the sourdough is of paramount importance. In rye sourdoughs some standards have been set in order to obtain an overall improvement in bread quality and production [1], but wheat flour is a different matter. Formerly, there has been great diversity among wheat varieties, quality of protein, and extraction rates, which affect breadmaking performance in much higher degree than in the case of rye flours. On the other hand, the absolute necessity for controlling enzymatic activity in rye flours does not apply to wheat. Recommendations point to regulation of acidity of bread to a value of 3.5–4 so to obtain a bread of pleasant flavor [43], and to increasing the levels of acetic acid in bread to delay mold spoilage, and thus, to reducing the use of preservative agents [32,44,45]. These goals can be attained only by our understanding and controlling the microorganisms involved in these processes.

II. INTERACTIONS OF MICROORGANISMS IN NATURAL SYSTEMS

The microflora fermenting the natural and commercial sourdoughs from the countries where they are commonly used have been discussed in other chapters. The great heterogeneity of the identified species of lactic acid bacteria and yeasts reflects the strong dependence on the steadiness of the processing conditions. It takes a well-controlled production to develop a constant and well-defined population coexisting for long periods of time. Depending on the type, genus, species, strain, and proportion of the microflora present in a sourdough system, it is possible to find indifference, antagonism, competition, or commensalistic association among the microorganisms for the available fermentable substrates [46]. Yeasts and lactic acid bacteria generally coexist because they do not compete for the same carbon source; the two microbial groups seem to establish a symbiotic relationship [24,47]. Competition for the substrate leads to consumption of the fermentable carbohydrates by yeasts, and greatly reduces the expression of the heterofermentative metabolism during leavening [24]. According to Ottogalli et al., [48], interactions between lactic acid bacteria and yeast associations in natural sourdoughs may lead to two situations: strict and weak associations.

Traditional sourdough processes usually do not rely on a fortuitous microflora, but on the use of starter sponges that are continuously propagated over long periods of time. The choice of suitable conditions ensures that microorganisms remain metabolically active through the process [49]. The fermentation conditions are strictly observed and repeated according to traditional schemes. Constant humidity and temperature and fixed times are used. Afterward the mother may be tightly packed to keep CO_2 and lower the redox potential [48]. These systems often develop strict associations characterized by few microbial forms of lactic rods and yeasts [48]. Only very few species of heterofermentative lactobacilli can be found in numbers that suggest a relevant contribution to fermentation [49]; among them the most frequent species are *Lactobacillus sanfranciscensis* [48,49] and *L. pontis* [49]. Yeasts belong preferentially to the species *Saccharomyces exiguus* and *Candida milleri* [48,49], and, in some occasions, to *S. cerevisiae* [49]. This comensalistic association is present in the San Francisco French sourdough bread, panettone, and other Italian baked products [3,4,47–49]; in them, *L. sanfranciscensis* and *S. exiguus* or *C. milleri* are the relevant microorganisms. Also, two strains of *L. sanfranciscensis* and one of *L. brevis* are the common population in a traditional German rye starter (Reinzuchtsauer) [50]. A specific combination of metabolic capabilities is suggested to be responsible for the prevalence of heterofermentative lactobacilli in sourdough [51]. They include the ability to ferment maltose and glucose producing high quantities of lactic and acetic acids [48] and the possibility of compensating the disadvantage of heterofermentative metabolism of generating only 1 mol of ATP per mole of glucose by the coupled action of a maltose-phosphorylase and the presence of proton acceptors (such as fructose, usual in sourdough systems), which increase ATP yields, compared with that of the standard homofermentative metabolism [51,52]. Selection of yeasts is based on their sensitivity and resistance to the unsaturated form of acetic acid produced by the lactic acid bacteria. *Saccharomyces cerevisiae* is very sensitive whereas *S. exiguus* and *T. holmii* are very resistant [53]. In starters where the bacterial population is composed of *L. brevis* ssp. *lindneri* and/or *L. sanfranciscensis* the addition of *S. cerevisiae* either commercial or maltose-grown, at levels of the order of those of the resident sourdough yeast, results in the elimination of baker's yeast after two transfers [54]. Generally, the characteristic combination is built with maltose-negative yeasts (*S. exiguus*, *C. milleri*, or *C. krusel*) [55,56] that are highly tolerant to acidic environment and release some amino acids (isoleucine, valine) essential for the lactic acid bacteria [48].

Weak associations characterize by a more changeable and heterogeneous population of yeasts like *Saccharomyces*, *Candida*, *Hansenula*, *Pichia*, and *Rhodotorula* in counts about 10^5–10^8 cfu/g, and lactic acid bacteria at about 10^7–10^9 cfu/g. Rods (10^8–10^9 cfu/g) belong to obligate homofermentative, facultative heterofermentative, and obligate heterofermentative lactobacilli, and cocci are from the genera *Pediococcus*, *Leuconostoc*, and *Enterococcus* (10^5–10^8 cfu/g) [48]. Their contemporary presence is related to a common origin and is conditioned by media resources and environmental conditions. Among the different groups a mutualistic relationship is not established; on the contrary, the competition for nutritional requirements selects continuously the strains that better fit in the operative conditions of each breadmaking. Besides the dominant strains, a collection of microbes that functions as an emergency pool in case of failure of the forefront microflora is also present. Fortuitous fermentation usually relies on homofermentative species [49]. In this selection the bacteriophages play an important role, and a real risk of consecutive viral contamination from a dough to another has been found. The changeability and unsteadiness of the microflora and, consequently the variability of the organoleptic properties of bread is a result of the artisan procedures as well as the use of *S. cerevisiae* to fasten the process. Examples of this association can be found in wheat sourdoughs from Italy [47,48,57,58], France [9,59], and Spain [7,8], as well as cracker doughs [5] and rye sourdoughs from Germany [1], Switzerland [17,18], and Finland [12]. On occasion, the initial community containing a multiform population develops through successive transfers toward a stricter ecosystem composed of fewer species [55,60].

Part of the problems in defining type of interactions arises from the difficulty in naming and identifying the lactobacilli isolated from sourdoughs. In the experience of Vogel et al. [61], *L. sanfranciscensis*, *L. brevis*, *L. fermentum*, and *L. fructivorans* are the predominant microorganisms in sourdough, but they do present many difficulties to be taxonomically identified. For example, strains initially identified as *L. brevis* belong to a new species designated *L. pontis* [56]. *L. sanfranciscensis* and *L. pontis* are obligate heterofermentative species, broadly differing in their metabolic potential [49] but living in close association, being very difficult to physically separate [61]. In turn, *L. sanfranciscensis* (usually found in wheat sourdoughs) [3,4,47] has been considered as synonymous [62,63] with *L. brevis* ssp. *lindneri*, a typical microorganism in rye sourdough [22,47,57,61,64]. According to Vogel et al. [61], *L. brevis* ssp. *lindneri* and *L. sanfranciscensis* are distinguished by the distinct amino acid in the interpeptide bridge of the peptideglican. *L. sanfranciscensis* contains L-alanine, whereas *L. brevis* ssp. *lindneri* contains D-aspartic acid.

The type of process greatly determines the evolution of microorganisms. Traditionally, a multistage breadmaking process involving a fresh sour, a basic sour, and a full sour is prepared [1,33,52]. The continuity of the microflora is ensured by consecutive reinoculation from a previous batch. Although assorted experience points to growth of yeast in fresh sour, promoting of lactic acid bacteria growth and acid production in basic sour, and acid formation and leavening in full sour [1,52], recent experience indicates simultaneous growth of lactic acid bacteria and yeasts in all the stages [52]. In dough-type sourdoughs, referred to as type I, made through these processes [52,65], *L. sanfranciscensis* and *L. pontis* are predominant and always found concomitantly with *C. milleri*, which contributes to the leavening of the dough.

Continuous production and long-term one-step production are the most predominant recently developed processes for sourdough preparation. Long-term fermentation results in "silo sours," referred to as type II. Under the conditions of these processes *L. sanfranciscensis* is not competitive; then the spectrum of microorganisms involves other genera and species, such as *L. pontis* and *L. reuteri* [53,65], whereas fermenting yeasts are not found in significant amounts. Type III sour doughs consist of powdered sours in which *L. plantarum*, *L. brevis*, and *Pediococcus pentosaceus* are the dominant strains [65].

In continuous processes temperature is lower and yeasts can develop in significant number [52,56]. Numerous species can survive in these processes [66] after successive sets of production, i.e., *L. fructivorans*, *L. delbrückii*, *L. cellobiosus*, *L. brevis*, *L. fermentum*, *L. plantarum*, and *L. brevis* ssp. *lindneri*, and the yeast *S. cerevisiae*.

III. METABOLIC FEATURES OF MICROORGANISMS AFFECTING TYPE OF INTERACTION

A. Metabolism of Carbohydrates

Free sugars and oligosaccharides (malto- and fructoglucans) are potential fermentable substrates for yeast and lactic acid bacteria in bread production [67–71]. The baker's yeast *S. cerevisiae* has saturable kinetics for hexoses and maltose (Fig. 1), and possess α-glucosidase and β-fructosidase [72].

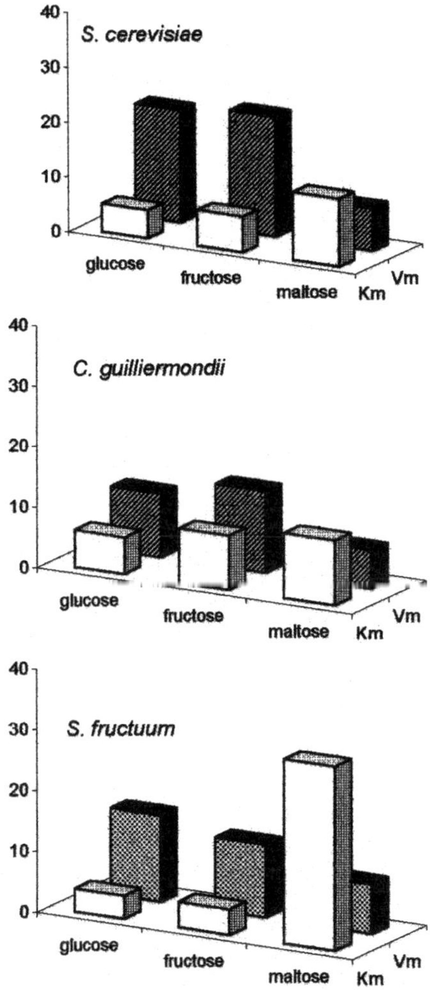

Figure 1 Kinetic parameters of sugar uptake in yeasts (K_m: mM; V_m: nmol/10^6 yeast.min). (From Ref. 72.)

Hexose transport occurs by an active metabolism bound mechanism, whereas the maltose transport system is inducible [73–78], the synthesis of the maltose permeases being the limitant step in maltose fermentation. Glucose and fructose repress the induction of maltose permeases, with glucose the more effective [77,78]. A preferential uptake of glucose over fructose has been described for *S. cerevisiae* [79,80], with rates of fermentation equal for both sugars when sucrose is the substrate. Two mechanisms of multiple sugar uptake [80] or glucose competitively inhibiting fructose uptake by the membrane carrier have been suggested [79]. *Saccharomyces cerevisiae* lacks α-amylases, but in dough systems, flour and/or exogenous amylases split starch molecules into fermentable sugars and malto-oligosaccharides [81]. *Saccharomyces cerevisiae* hydrolyzes maltotriose, although trisaccharides are fermented to a lesser extent [70,82]. The yeast does not ferment malto-oligosaccharides with degree of polymerization (DP) 4 to 8, nor did the presence of glucose stimulate their fermentation, probably due to a lack of the appropriate transport system for these oligosaccharides [82]. The extracellular yeast invertase, extremely active on sucrose, also splits the linkages in fructans; the rate at which a specific type of fructan is hydrolyzed decreases with increasing DP. Depending on fermentation time, most fructans with DP3 to DP5 are hydrolyzed, representing about 50% of the fructans in white flour, into fructose and glucose [83]. Other yeast species such as *S. fructuum*, *C. boidinii*, and *C. guilliermondii* have similar kinetics for hexose transport to that of *S. cerevisiae* [72], indicating that high-affinity systems are active for these sugars. Values of K_m for maltose are greater than for hexose, and they agree with the fermentation power shown by these yeasts. The sourdough yeasts *C. milleri* and *S. exiguus* do not utilize maltose [84].

Sugars used by lactic acid bacteria as energy sources vary within species and even within strains. The most common lactic acid bacteria identified in sourdoughs are capable of fermenting pentoses [68], hexoses, and the disaccharides sucrose and maltose [68,72,85], although some species, such as *L. sanfranciscensis*, are specific for maltose [84]. The capacity of lactic acid bacteria to convert starch to lactic acid is not a common characteristic among lactic acid bacteria. Amylolytic activity has been described in other foods for some strains of *L. acidophilus, L. cellobiosus*, and *L. plantarum*, species also common in sourdoughs [86]. In *L. plantarum* this property was extracellular, and the enzyme was synthesized when glucose concentration descended to low levels [86]. Within different strains of sourdough lactobacilli screened for this activity, *L. sanfranciscensis* showed the lowest α-amylase activity and *L. plantarum* the highest (Fig. 2) [81]. Being lactobacilli inoculated at about 10^9 cfu/100 g flour in dough systems, flour is the main quantitative contributor to the overall α-amylase activity [81]. However qualitative differences can be produced due to the specificity of the enzyme for low molecular weight substrates [87–89].

Strains of *L. brevis, L. plantarum, L. cellobiosus*, and *Enterococcus faecium* isolated from wheat sourdoughs have saturable kinetics for hexoses, except *L. brevis*, which follows nonsaturable kinetics for glucose incorporation (Fig. 3). Sucrose and maltose incorporation is linear for all strains even

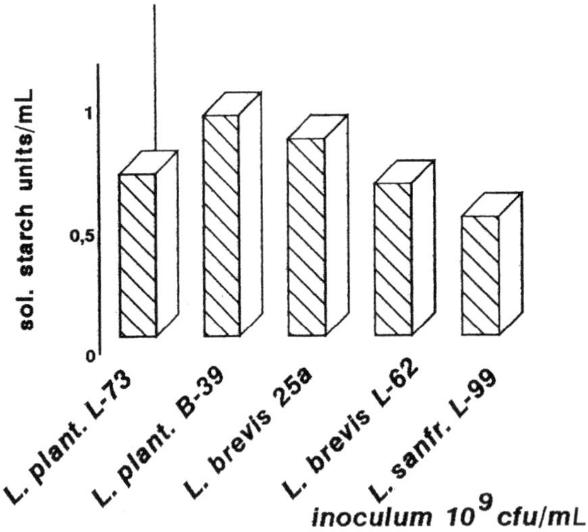

Figure 2 Amylolytic activity of some lactic acid bacteria used in sourdough processes. (From Ref. 81.)

when the external concentration is increased to 1 mM (Fig. 3), and all have α-glucosidase and β-fructosidase [72]. In these strains the maltose transport uses the phosphotransferase-phosphoenolpyruvate-dependent system (PTS-PEP system) [72]. *L. sanfranciscensis, L. pontis, L. reuteri*, and *L. fermentum* have maltose phosphorylase, a key enzyme for the predominance of these microorganisms in sourdough [49,56,90]. The transport of maltose is mediated by a H^+/maltose symport system, and a maltose-inducible uniport system is involved in glucose export [49,51,90]. In these microorganisms, maltose utilization is very effective and not subjected to glucose repression. *Lactobacillus sanfranciscensis* hydrolyzes maltose [49,51,90–92] and accumulates glucose in the medium in a molar ratio of about 1:1. Glucose-1-phosphate is metabolized while glucose is excreted to avoid excessive intracellular concentration. Excretion of glucose takes place in concomitance with maltose availability and after maltose has been depleted the consumption of glucose begins. The glucose excreted may be used by maltose-negative yeasts, or may prevent competitions from utilizing abundant maltose by glucose repression, thereby giving an ecological advantage to *L. sanfranciscensis*.

Some lactic acid bacteria common in sourdough systems are fructose negative (*L. brevis* ssp. *lindneri, L. sanfranciscensis*) and grow faster in maltose than in glucose; *L. plantarum* prefers maltose and glucose over

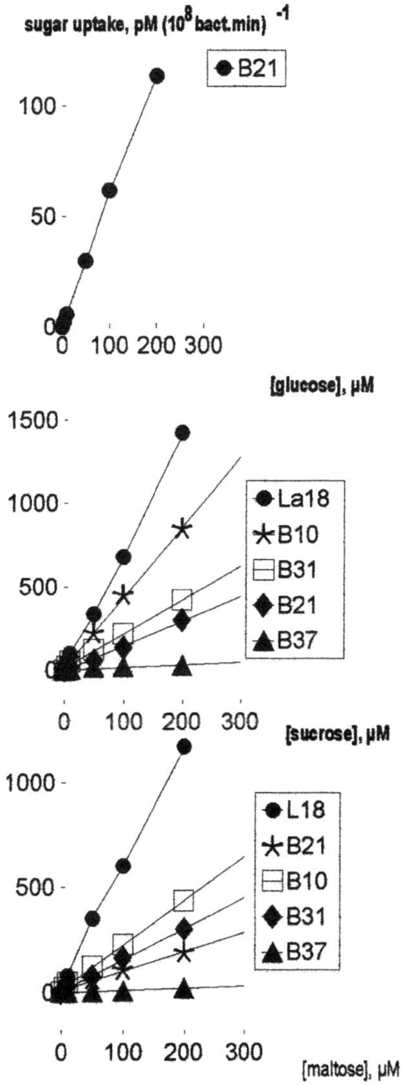

Figure 3 Sugar uptake by lactic acid bacteria (B31, La18: *L. plantarum*; B-21: *L. brevis*; B10: *Leuconostoc mesenteroides*; B-37: *L. cellobiosus*). (From Ref. 72.)

fructose for rapid growth and weakly ferment sucrose [57]. Cofermentation enables microorganisms to use substrates that are otherwise nonfermentable and increase microorganism adaptability to difficult ecosystems; *L. sanfrancisco* and *L. brevis* ssp. *lindneri* are examples of that [51,91]. Yeast invertase on sucrose liberates glucose and fructose, and enables partial growth of the lactobacilli, which leads to mutualism in sourdough fermentation [91]. They are able to coferment fructose in the presence of glucose or maltose [51,91], increasing cell yield, lactic and acetic acid production, and synthesis of mannitol, while decreasing the levels of ethanol. The utilization of fructose depends on the continuous presence of maltose in the medium, and the two carbohydrate are consumed in a molar ratio of about 2:1 (93). A cometabolism of maltose or glucose and citrate has also been observed in citrate-negative strains [51,91,94].

In flour-water brews, *L. brevis* utilized glucose and maltose at a faster rate than two strains of *L. plantarum*, which presented noticeable differences between strains, but the homofermentative strains were more efficient in consuming fructose [81]. The amylolytic activity expressed by *L. plantarum* (Fig. 2) increases the amount of maltotriose over that of flour, reaching a maximum after 24 h, this being the only trisaccharide metabolized, and only by *L. brevis* from several *Lactobacillus* species. Larger oligosaccharides with DP up to 7 accumulate during the last stages of fermentation, and are further degraded by enzymatic reactions to smaller fragments, which depended on the flour and the strain of *Lactobacillius* [81].

Heterofermentative lactobacilli such as *L. sanfranciscensis, L. brevis, L. bucheneri*, and *L. fermenti* common in sourdoughs are stimulated by oxygen [26,32,44,49,51]. Oxygen shifts the metabolic pathway from the ethanol to the acetate route, at the same time doubling the amount of ATP formed by 1 mol of glucose; the resulting increase in energy yield enhances growth rate and growth yield [32,44,51,95]. Increases in acetic acid production due to oxygenation in wheat flour systems range from 54% (*L. brevis*) to 269% (*L. sanfranciscensis*) (Fig. 4) [32]. The heterofermentative *L. pontis* requires anaerobic conditions for growth [49]. Agitation of cultures with *L. pontis, L. reuteri, L. fermentum*, and *L. amylovorus* [56] does not affect the log phase, but cells generally grew better and the acetate/ethanol ratio did not change. Homofermentative strains such as *L. plantarum* are also affected by oxygenation [96]. *Lactobacillus plantarum* grown aerobically on glucose consumes molecular oxygen with either glucose, lactate, or pyruvate as substrate [95,97,98]. Molecular oxygen is reduced to H_2O, the reduction proceeding via oxygen peroxide, involving an NADH oxidase and an NADH peroxidase. Several mechanisms are available, and the final byproducts will depend on the specific substrate utilized [95,97]. Accumula-

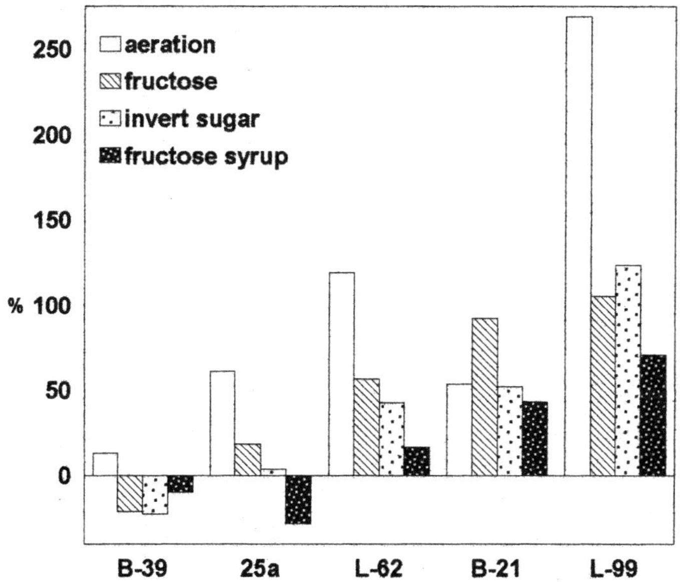

Figure 4 Effect of aeration and fructose on the production of acetic acid by lactobacilli in sour doughs (expressed as a percentage of activity shown by control samples). (Fermentation conditions: 44 h, 35 °C, consistency 350 g dough/100 g flour, B-39: *L. plantarum*, 25a, L-62 and B-21: *L. brevis*, L-99, *L. sanfranciscensis*). (From Ref. 32.)

tion of hydrogen peroxide was always preceded by an increase in the rate of oxygen consumption [98,99] and accompanied by a switch in metabolism that resulted in acetate rather than lactate accumulation in aerobic cultures [96,99]. Oxygen also promotes an increase in lactic acid production in homo- and heterofermentative lactobacilli, but increases are lower than those observed for acetic acid (7–32%) [32,44]. Aeration of sourdoughs containing homo- and heterofermentative lactobacilli seems to induce an activation of the energetic metabolism in addition to favoring the formation of acetic acid [95].

Proton acceptors, such as fructose usually found in flour, partially push the metabolism of heterofermentative lactic acid bacteria toward the acetate kinase pathway, producing traces of mannitol and increase of acetic acid. Fructose is required for growth of *L. pontis, L. reuteri, L. fermentum*, and *L. amylovorus* [56]; it is used in part to produce mannitol and acetate, and in part to be metabolized to lactate and ethanol. Maltose plus fructose reduces the log phase for *L. pontis* when compared with maltose alone, and increases growth rates and final cell yield. Fructose promotes an increase in cell growth, acetic acid production (from 3.3 to 15 mmol/kg), and fermentation quotient with *L. brevis* ssp. *lindneri*; it also decreases ethanol production, and although gas production is slightly modified, the gas production rate is almost doubled [100]. Increases in acetic acid production have also described for *L. brevis* and *L. sanfranciscensis* [32,44,49] but the overall effects (increases of 92–123%) are smaller than those observed with oxygenation [32,44]. The efficiency of fructose as proton acceptor depends on concentration, temperature, and dough consistency [45], as well as in the form of the preparation; pure fructose is more effective than invert sugar or fructose syrup [32], probably due to osmotic effects. The addition of fructose results in no effect or a negative effect (ranging from – 10% to – 50%) on lactic acid production [32,44].

Carbohydrate metabolism of lactic acid bacteria and their energy yield are greatly influenced by the associated yeasts and vary according to the type of sugar [57,91,101]. When *L. sanfranciscensis, L. brevis* ssp. *lindneri* or *L. plantarum* are associated to *S. exiguus* (maltose negative), the lactobacilli completely take up maltose and bacterial cell yield increases (Fig. 5) [100]. In association with *S. cerevisiae* a decrease in bacterial metabolism occurs due to the faster consumption of maltose and especially glucose by the yeast, which reduces the availability of glucose when both microorganisms grow together. Confirming this ethanol synthesis increases. This can create critical situations if there is not another source of fermentable carbohydrates. The

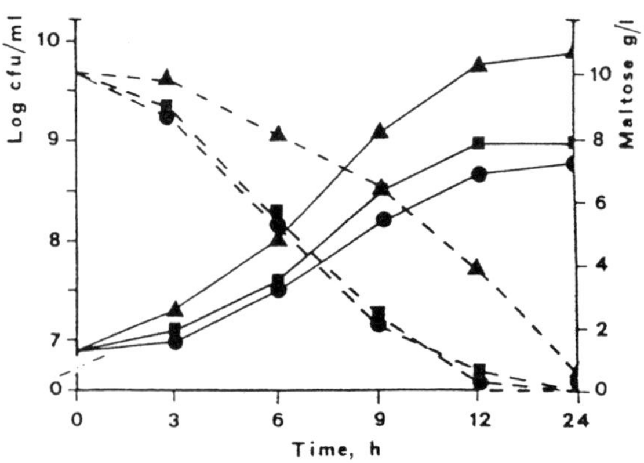

Figure 5 Evolution of maltose (dotted line) and bacterial growth (solid line) in monocultures of *L. brevis* ssp. *lindneri* (●) and in cocultures with *S. exiguus* M14 (▲) or *S. cerevisiae* 141 (■) in medium containing maltose. (From Ref. 103.)

addition of fructose increases acetic acid concentration, whereas the addition of maltose raises the amount of lactic acid. Fructose does not affect the growth of *S. cerevisiae* but improves gas and ethanol production. Fructose when added to associations of microorganisms increases ethanol production, accelerates starter activity, and ensures the appropriated lactic acid bacteria fermentation in terms of pH, cell numbers, and lactic and acetic acid production [100]. These observations are also valid for *L. plantarum* and its associations with glucose or fructose as carbon source. With yeast present, *L. brevis* ssp. *lindneri* increases cell yield only slightly in the presence of sucrose, whereas *L. plantarum*, which is capable of using both hexoses resulting from yeast invertase splitting, is strongly stimulated [57]. The presence of baker's yeast also influences the amounts of soluble dextrins produced during sourdough fermentation. Dextrins with DP 5 to 7 accumulate to a greater extent in yeasted sourdoughs than in these containing only lactobacilli, whereas maltotriose and maltotetraose reach higher values in the latter case [71].

B. Production/Utilization of Nitrogen Compounds

Free amino acids are growth factors for fermentative microorganisms. In culture media, glutamic acid, isoleucine, and valine are essential for growth of *L. brevis* ssp. *lindneri* or *L. plantarum*; arginine, methionine, and leucine are stimulatory; and other amino acids have no effect [102]. Each individual amino acid, except lysine, cysteine, and hystidine, is suitable for growth of *S. cerevisiae* or *S. exiguus* [62,102,103]. Cocultures of these strains with either yeast enhance the growth of the lactobacilli. The stimulation seems to be due to the partial competition for the nitrogen sources, since in a mixture of ammonium salts and amino acids the yeast preferentially use the former. The same trends have been found with gluten hydrolysates as medium [102]. Yeast also excretes amino acids during growth due to a change in membrane permeability or as a consequence of accelerated autolysis induced by differences in osmotic pressure between the medium and the yeast cell [102]. The biomass of *S. cerevisiae* releases γ-aminobutyric acid, proline, valine, lysine, isoleucine, glycine, and alanine, corresponding to about 5% of amino acid content of bread doughs, whereas amino acids excreted by biomass of *L. brevis, L. plantarum*, or *E. faecium* are mainly glycine and alanine and represent about 1–2% of that supplied by flour. Contribution of yeasts and lactic acid bacteria to the peptide content of the bread dough is greater than that of amino acid (32% and 20%, respectively) [104,105].

Release of amino acids due to proteolytic activity of yeasts is unlikely to occur [103], but this activity is common among lactic acid bacteria. In general, it suffices to provide the necessary amino acids for growth [106].

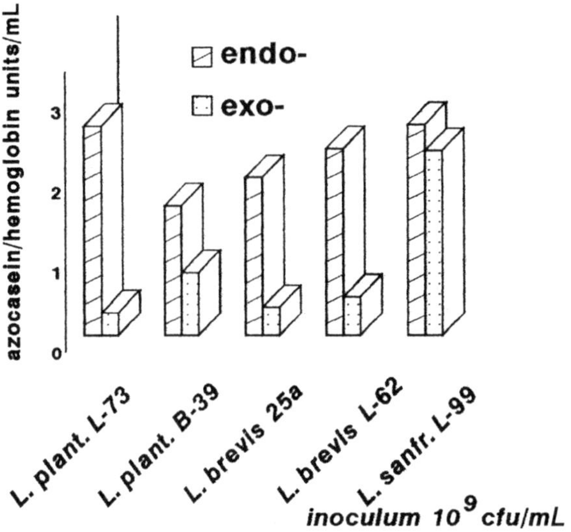

Figure 6 Endo- and exoproteolytic activity of some lactic acid bacteria used in sourdough processes. (From Ref. 110.)

Proteolytic activity is a quality inherent to strain not to species, often being of exopeptidase character, liberating low molecular weight compounds [107]. Extracellular proteinases are less common [108] and usually have a degree of specificity [106]. Differences in endoproteolytic activities of strains of *L. brevis, L. plantarum*, and *L. sanfranciscensis* isolated from sourdoughs are small (Fig. 6) [109]; however, estimated protein chain length decreased during 24 h fermentation with *L. plantarum* and *L. brevis*, indicating endoproteolysis upon proteins and polypeptides [110]. Exoprotease action depended on pH, with *L. sanfranciscensis* showing higher exoproteolytic activity than *L. brevis* and *L. plantarum* (Fig. 6) [109]. The proteolytic activity of some strains of *L. brevis* ssp. *lindneri* and *L. plantarum* has shown to be very similar [103]. The participation of lactic acid bacteria to the overall proteolysis of a breadmaking process is considered limited by some authors [104,111], whereas others esteem that lactic acid bacteria are responsible for about two-thirds of the proteolytic activity of sourdough [112,113]. In any case, they induce changes due to differences in specificity towards the substrate [109]. *Lactobacillus sanfranciscensis* shows the highest aminopeptidase, dipeptidase, tripeptidase, and iminopeptidase activities when compared with other lactic acid bacteria strains, which can be important in wheat flour that contains appreciable quantities of low molecular weight peptides [91]. The possibility of generating bitter peptides also has to be considered [108].

The presence or absence of yeast affects peptide and amino acid metabolism. In the presence of yeast, there is a stronger demand on amino acids and low molecular weight peptides during fermentation [114]. At the beginning of fermentation, proteolytic activity is low due to nonappropriated pH, and generation of amino acid and peptides is small; at this stage of the process, the yeasts follow a log phase of growth that induces a strong demand for nitrogen. Lactic acid bacteria also have a long lag phase and develop metabolic activities at a slow rate during the first 4 h. Thus, accumulation of amino acids, as a consequence of the hydrolyzing action of proteases from lactic acid bacteria and flour, is not observed until the later stages of fermentation [115].

Predominant amino acids in nonfermented wheat sourdoughs are dicarboxylic acids and amides, and aliphatic amino acids, whereas hydroxyl- and sulfur-containing amino acids are minor components [116]. An increase in the levels of different amino acid during sourdough fermentation has been widely stated [24,113,114,116]. In rye sourdoughs, *L. plantarum* shows higher generation of amino acids than *L. brevis* ssp. *lindneri* or *L. fructivorans* [113, 114]. In wheat sourdoughs *L. brevis* ssp. *lindneri*, *L. sanfranciscensis*, *L. brevis*, and *L. plantarum* increase the level of aliphatic, dicarboxylic, and hydroxy amino acids (Fig. 7) [24,103,116], and all aliphatic and dicarboxylic amino acids are essential for amino acid metabolism [24,104]. The yeasts *S. cerevisiae* and *S. exiguus* decrease the levels of total L- and D-amino acids in a similar way, differing only in aliphatic amino acid composition [62,117]. The rate of removal depends on the strain; thus, *S. cerevisiae* is more effective than *C. krusei* in removing the amino acids from the dough [113,114]. Combinations of yeasts and lactic acid bacteria show intermediate values for total amino acid levels. *Saccharomyces cerevisiae* and *L. brevis* ssp. *lindneri* produce sourdoughs with lower total amino acid than yeasts alone, with a marked decrease in aliphatic, dicaboxylic, and hydroxy amino acids [62,103,116]. Utilization of amino acids leads to an increase in yeast counts. Similar trends occur for *S. exiguus* and *L. brevis* ssp. *lindneri* and combinations of each yeast with *L. plantarum* [103]. The L isomer is the usual form of amino acids produced by microorganisms, but the D isomer appears in greater amounts in the presence of starters than in their absence, giving lactic acid bacteria greater concentrations than yeasts. The possible biotic generation of D forms was also confirmed in associations of lactic acid bacteria and yeasts. The cellular excretion of amino acid could provide L and D forms, but a selective utilization of the L isomers will lead to accumulation of the D form in fermented doughs [117].

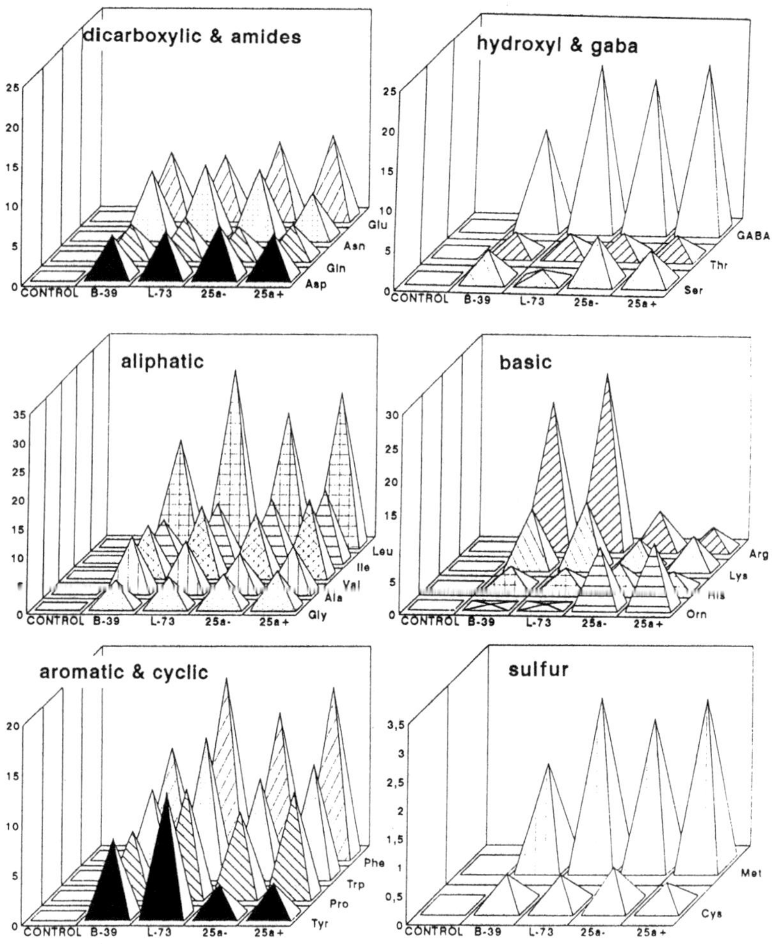

Figure 7 Influence of lactobacilli and yeast on individual amino acid profile of 22 h fermented sourdoughs. (B-39 and L-73: *L. plantarum*; 25a: *L. brevis* −/+ indicates the absence or presence of *S. cerevisiae*, respectively.) (From Ref. 117.)

The estimated free peptide content of sourdoughs with either *L. brevis* or *L. plantarum* is lower than the estimated amount of amino acids. Changes in peptide the first 4 h of sourdough fermentation are small. The relative rate of change in amino acids and peptides, particularly in the case of *L. brevis*, shows the higher reactivity of peptides with respect to amino acids during fermentation [110]. Both lactobacilli induce a net decrease in peptides with

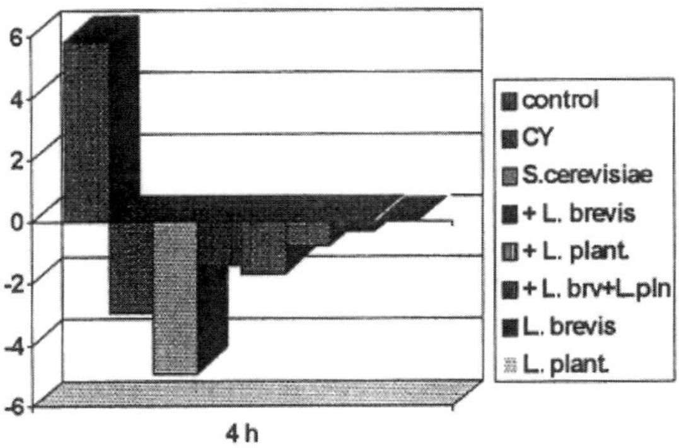

Figure 8 Influence of lactobacilli and yeasts on the amount of low molecular peptides (< 10,000 Da) of sourdoughs after 4 h fermentation. (From Ref. 116.)

molecular weight < 10 kD (Fig. 8). Metabolism of peptides is greater for pure cultures of *S. cerevisiae* than for compressed yeast, and their removal undergoes a further increase in sourdoughs containing mixed cutures of *S. cerevisiae* and lactobacilli [115].

After proofing, doughs with *L. brevis* or *L. plantarum* show a significant decline in dicarboxylic acids and amides, serine, and hydrophobic amino acids as well as in total amino acid content. In general, homofermentative strains induce bigger decreases. Predominant amino acids such as alanine and γ-aminobutyric acid are sharply promoted as a consequence of enzymatic reactions [116]. Baking decreases total amino acids in about 40–75%, without any particular difference among starters of different composition. The reduction is more pronounced for proline, asparagine, leucine, glutamic acid, and lysine, and is positively related to the aromatic characteristics of breads [62].

Pure cultures of *S. cerevisie* induce a net fall in several salt (SSP)–soluble proteins and an increase in others, in particular the 53,000-Da subunit; the degrading activity also extends to numerous subunits of dioxane (DSP)–soluble proteins [118]. Some proteolytic activity has been evidenced in some strains of yeasts [119]. The use of compressed yeast results in minor effects on these protein fractions. Homofermentative lactobacilli increased the level of SSP in mixed cultures, whereas *L. brevis* induced increase of some SSP and DSP subunit and decrease of others. The effect is strongly dependent on the composition of the starters [118].

Metabolism of amino acids and proteolytic activities depends on the microbial starter [112,114,120,121] and on processing conditions [121–123]. In wheat sours, the extraction rate of flour and the fermentation temperature are reported to be the main factors with positive influence on the level of total free amino acids [123] and on the extent of accumulation of hydrophobic and basic amino acids during fermentation [122]. The kind of flour also influences proteolysis; rye flours with higher ash content lead to higher levels of amino acids, particularly leucine, alanine, asparagine, lysine, and proline [112]. Dough yield also influences total amino acid content, mainly in the presence of yeast; soft sourdoughs contain lower amount of amino acids [121].

IV. FUNCTIONAL AND QUALITY RELATED ASPECTS OF INTERACTIONS OF MICROORGANISMS IN DOUGH FERMENTATION

Compatibility of associations between lactic acid bacteria and yeasts and their effects on dough leavening, acidity, and rheology, as well as on bread quality, is widely documented. *Saccharomuyces cerevisiae* follows different fermentation kinetics than other sourdough yeasts such as *S. exiguus* [55,100], *S. fructuum*, or *C. boidinii* [101,124]. Sourdough yeasts have lower fermentative activity and all of the CO_2 produced is retained in the dough; the absence of porosity explains the unsatisfactory bread crumb texture [100]. Heterofermentative lactobacilli produce a small gas release [1,55,101,124] that for *L. brevis* ssp. *lindneri* represents about 22% of that of *S. cerevisiae* [100]. Thereof they give breads with flat volume and coarse crust and crumb. Homofermentative strains do not have leavening ability and are not suitable for baking [124]. The acidification properties of lactic acid bacteria are variable [1,85] and, in general, heterofermentative strains give greater acidity than homofermentative strains [35]. The association of *S. cerevisiae* and *L. brevis* ssp. *lindneri* shortens the time to reach the maximal CO_2 and ethanol evolved, but it does not modify the amount generated. The decrease in pH due to acid production is responsible for an initial activation of yeast metabolism. *Saccharomyces cerevisiae* strongly affects carbohydrate availability, causing a reduction in *L. brevis* ssp. *lindneri* fermentation and therefore decreasing the production of acids and cell counts [100]. The association with *L. plantarum* improves *S. cerevisiae* fermentation [125–127], demonstrating the adaptability of the yeast to grow in association with homofermentative strains, but with *L. brevis* the bacterial cell yield increases (Fig. 9) [126]. Association of *C. milleri* (maltose negative) with *L. plantarum* also promotes leavening but acidification falls down, indicating that both cultures are not compatible with each other; with *L. brevis*, the opposite effects are observed (Fig. 9) [126]. The addition of *L. brevis* ssp. *lindneri* to *S. exiguus* results in a reinforcement of fermentation

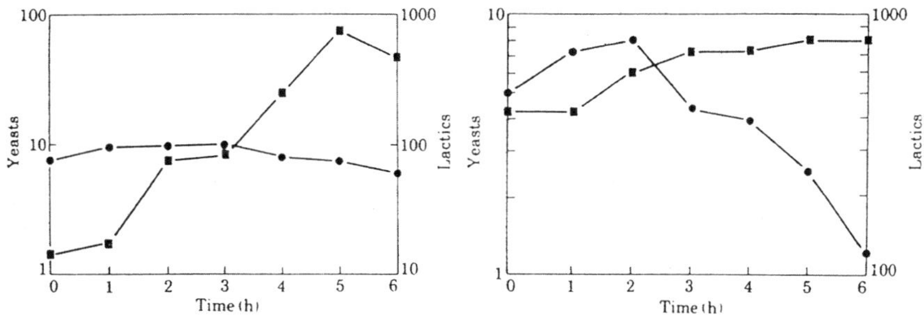

Figure 9 Changes in microbial profiles in sourdoughs containing *L. brevis* (■) and the yeasts *S. cerevisiae* (● left) or *C. milleri* (● right). (From Ref. 126.)

for both strains, probably due to the lack of competition for the substrate. The positive effect on gas and ethanol production gives loaves of satisfactory texture. *Lactobacillus plantarum* when used with *S. exiguus* does not cause further effects to those mentioned for the yeast alone [126].

In associations of *S. cerevisiae* and *L. plantarum* or *L. brevis*, the latter leads to lower pH, higher total titratable acidity (TTA), and higher organic acids level (mainly lactic acid) in doughs and breads than the homofermentative strains (Fig. 10) [1,37]. The presence of yeast in the starter reduces acidification in doughs and breads [16,37]. Sour doughs with homofermentative lactobacilli have lower acetic acid production (about 16–18%) and variable lactic acid production (−26% -L-73-, +43% -B-39-) when the yeast is included (Table 1) [37,128]. For *L. brevis*, the presence of yeast increases acetic acid content by 66%. With both lactobacilli loaves have uniform grain and good volume. Heterofermentative strains have stronger acid and acetic taste, and variable lactic taste. Sensory acid and acetic taste are correlated with TTA and acetic acid content, but not with lactic acid content. They also have greater resistance to mold contamination due to the higher acetic acid concentration in bread [37].

The addition of yeast to sourdough gives loaves of greater volume and slice height. Crumb firmness is, in general, inversely related to bread volume; however, *L. brevis* gives harder bread crumb for similar bread volumes than *L. plantarum*, mainly in the absence of yeast at the sourdough stage (Table 1) [37]. The yeast also influences organoleptic properties of breads; stronger flavor was observed in breads containing yeast in the sourdough. A strong but satisfactory and characteristic pleasant flavor corresponds to samples with acidity values from 2.8 to 4.34 [37], a broader acidity interval than is recommended for wheat breads [43]. Atypical flavor is generally related with a low percentage ($\leqslant 10\%$) of sourdough when homofermentative starters are

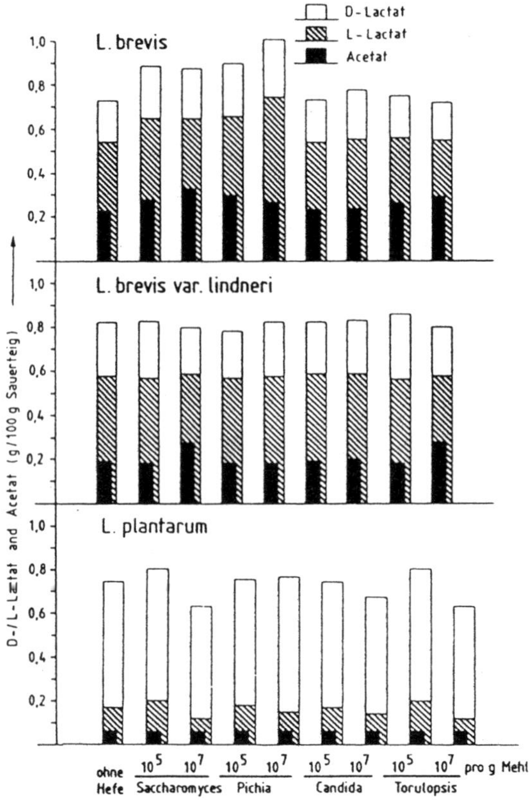

Figure 10 Influence of yeasts on the production of lactic and acetic acids by lactobacilli in sourdoughs. (From Ref. 1.)

used, and with a high percentage (>17.5%) of sourdough addition in heterofermentative samples [37].

Bacterial starters affect dough rheology (Table 2). Sour doughs with homofermentative lactobacilli and yeast used at 17.5% of addition give the softest doughs with longest development times in the farinograph. With *L brevis*, increments in sourdough addition regularly increase development time and do not modify dough consistency. Dough stability and degree of softening depend on both the presence of yeast and the amount of sourdough. In the extensigraph, the increase in the percentage of sourdough with *L. plantarum* decreases the maximal resistance to the extension and the energy, whereas with *L. brevis* the contrary effect is observed. The presence of yeast in the sourdough leads to further decreases in energy [37].

Table 1 Effects of *Lactobacillus*, Yeasts, and Percentage of Addition of Sourdough on Physicochemical Properties of Breads

Bacterial culture	Yeast	% sourdough	pH	TTA, mL 0.1 N NaOH/10 g	Lactic acid g/100 g as is	Fermentation quotient	Volume (mL)	Texture (g)
L. plantarum	None	10	5.54	2,97	0.0808	2.39	1990	5270
		17,5	5.07	3,58	0.1842	5.58	2390	3715
		25	4.72	4,73	0.3299	10.68	2430	3790
	S. cerevisiae	10	5.50	3,13	0.0900	3.61	2550	2725
		17,5	5.18	3,71	0.2059	8.12	2990	1106
		25	5.01	4,25	0.2554	10.38	2810	1850
L. brevis	None	10	5.03	3,45	0.2441	13.56	2490	4400
		17,5	4.76	4,23	0.3754	8.45	2260	5330
		25	4.49	5,11	0.6477	42.75	2060	9720
	S. cerevisiae	10	5.06	3,55	0.2613	14.52	2780	2550
		17,5	4.71	4,42	0.5979	17.95	2040	5606
		25	4.49	5,19	0.6941	18.96	2160	5904

Source: Ref. 37.

Table 2 Effects of *Lactobacillus*, Yeasts, and Percentage of Addition cf Sourdough on Rheological Properties of Bread Doughs

			Farinograph			Extenigraph		Maturograph		
Bacterial culture	Yeast	% Sourdough	Development time (min)	Stability (min)	Softening 10 min (BU)	Maximum resist (BU)	Energy, (cm^2)	Fermentation time (min)	Stability (min)	Dough level (BU)
L. plantarum	None	10	4	9	30	257	36,3	78	15	298
		17,5	4,5	3,5	120	205	30,5	77	28	255
		25	4	2,5	130	210	27,5	58	13	265
	S. cerevisiae	10	4	9	20	287	34,4	67	8	515
		17,5	5,5	5,5	60	230	23	54	8	570
		25	3,5	4,5	110	185	16,8	46	8	360
L. brevis	None	10	2,5	2,5	190	200	31	92	24	240
		17,5	3	4	140	233	33,3	72	12	265
		25	5	3,5	50	260	36,3	69	6	350
	S. cerevisiae	10	3,5	2,5	250	208	21,9	70	18	273
		17,5	3,5	5	130	235	32,2	74	12	350
		25	4,5	7,5	55	285	30,9	51	8	590

BU, Brabender units.
Source: Ref. 37.

Some strains, such as *C. boidinii* and/or *Entersisccus faecium* in association with *S. cerevisiae* and/or *L. plantarum* and *L. brevis*, significantly increase the amount of acetic and lactic acids produced during proving. The combination of *L. plantarum* and *Ent. faecium* increases the proportion of the *L*(+)-lactic isomer over that produced by *L. plantarum* alone [101]. In general, except for *S. cerevisiae*, an increase in the proportion of yeast or lactic acid bacteria does not improve the performance of the microorganism used, but mixtures of *S. cerevisiae* and the yeasts *C. boidinii*, *C. guilliermondii*, and *S. fructuum* improve their baking performance when lactic acid bacteria are also included [124].

Associations of *L. sanfrancisco* and *L. plantarum*, strains differing in acetic acid production, acidification level and demand for nutritional requirements [15] correspond to the commensalism class. The associated population has cell yields and physiological activity exceeding that of the monocultures. However, cultures of *L. plantarum* combined with *L. delbrueckii* or *L. leichmanii*, aimed at starting soda crackers, have fermentation rates similar to those of individual cultures, presenting a type of indifferent association [129].

The proliferation and fermentative activity of the sourdough microflora can be controlled by appropriate adjustment of fermentation time and temperature, and sourdough consistency [28,30]. Sourdoughs with low consistency produce more lactic acid [28,31] and ethanol, ferment faster, and consume more fermentable sugars [31]. In wheat flour doughs, extraction rate is a key factor to be considered [16,29–31]. The higher ash content as well as the swelling and buffering capacity of high extraction rate flours affects the activity of the microorganisms present [29,30]. On the other hand, some associations, including strains of *S. cerevisiae*, *L. brevis*, and/or *L. plantarum*, in straight or sourdough processes work better in rye than in wheat flour systems [130].

V. FACTORS IN THE FORMATION OF VOLATILE AROMA COMPOUNDS

Homofermentative lactobacilli characterize by the production of ethyl acetate, acetaldehyde, hexanal, and diacetyl. The strains of *L. plantarum* have the most homogeneous behavior within several homofermentative species. Heterofermentative strains distinguish by nonanal, ethyl acetate, and octanal. In this case, *L. brevis* and *L. fementum* are the species presenting greatest variability within strains. The presence of aldehydes among the discriminating variables suggest than these compounds are not only derived from the enzymatic oxidation or autoxidation of the lipidic fraction of wheat but are also, in part, produced through the heterofermentative metabolism [131]. Isoalcohols (2-methyl-1-propanol, 2,3-methyl-1-butanol), with their respective aldehydes and ethyl acetate, are the characteristic volatile compounds originated by yeast in the sourdough. Oxidative yeasts (*C. krusei* and *C. norvegensis*) do not produce ethanol but increased hexanal, octanal, nonanal, and hexane contents. Considering all the strains belonging to each metabolism as a group, differentiation among species is mainly related to 2-methyl-1-propanol and 2,3-methyl-1-butanol (yeast activity), diacetyl and hexanal (homofermentative metabolism), and ethyl acetate (heterofermentative metabolism). Combinations of different lactic acid bacteria result in a balanced profile, corresponding to the integration of homo- and heterofermentative volatile compounds, and in a reduction in the maximal amounts observed when they are used individually. Association of lactic acid bacteria and yeasts result in a synergistic increase of alcohols with respect to that of *S. cerevisiae* alone. The main bacterial volatile components, ethyl acetate and carbonyls, significantly decrease. The effect is probably due to the combination of bacterial acidification and proteolysis producing amino acids and afterward isoalcohols. Combinations containing *S. exiguus* instead of *S. cerevisiae* presented similar effects when *L. brevis* is used, but *L. plantarum* has a detrimental effect in the synthesis of alcohol by the yeast [131].

Generally, *S. cerevisiae* produces more volatile components in the headspace vapor than *C. guilliermondii* or *C. boidinii* [132,133], leading glucose and sucrose to a greater variety of volatile compounds than maltose or flour [133]. *Lactobacillus brevis* and *Lactobacillus plantarum* produce a lower number of volatile components than yeasts, and flour induces the formation of more volatile components than sugars in the headspace fraction [133].

Associations of *L. brevis* ssp. *lindneri*, or *L. plantarum*, and *S. cerevisiae* increase the formation of yeast fermentation products (1-propanol, 2-methyl-1-propanol, 3-methyl-1-butanol) and the number of aroma components detected [62]. *Lactobacillus* and *S. exiguus* give a profile corresponding to that of bacterial activity. Ethanol and lactic and acetic acids differed for both lactobacilli. The association of *L. sanfranciscensis*, *L. plantarum*, and *S. cerevisiae* is common in wheat sourdough [57,125,134], and it has been the choice to guarantee an equilibrated aroma of wheat sourdough bread [41,62,134].

Different fermentation parameters (dough consistency, temperature, time) affect in a similar way the volatile composition of the sourdough produced by different mixed starters. Raising temperature from 25 °C to 30°C increases yeast fermentation; at 25 °C it characterizes by the formation of ethyl acetate, acetic and lactic acid, whereas at 30 °C ethanol, 1-propanol,

2-methyl-1-propanol and 3-methyl-1-butanol are the typical products; increasing the temperature to 35 °C does not modify the profile. Firmness of dough guarantees a greater contribution of lactobacilli to the volatile profile [62]. In soft doughs, ethanol and ethyl acetate are predominant, with high levels of isoalcohols in doughs containing homofermentative strains; these components are considered a consequence of yeast activity that developed more in presence of homofermentative species due to the lower concentrations of acetic acid generated. Tough doughs differ in ester composition; ethyl acetate in heterofermentative sour doughs and carbonyls in the homofermentative ones are more abundant than in softer doughs [41]. High ash content considerably promotes the overall concentration of volatile components, but there is no difference in qualitative composition; the greater amount of precursors in dark flours may account for the increase [62]. High extraction rate flours increase the amount of ethyl acetate and ethanol when inoculated with heterofermentative lactobacilli, but white flours have higher levels of other alcohols [134,135]. This effect on the concentration of ethanol and acids may be attributed to the buffering capacity of flours and to the presence of compounds such as phytic acid, which prolongs lactic fermentation in spite of low pH values.

Lactic acid bacteria, when used individually, did not produce short-chain (C_3–C_6) volatile organic acids, but when strains *L. plantarum* and/or *Ent. faecium* are combined with *S. cerevisiae* or *C. boidinii*, an increase in propionic and isobutyric acids and a decrease in isovaleric acid concentration are observed in relation to the amounts produced by either yeast alone [132].

After baking the amount of volatile components decreases. For *L. brevis* ssp. *lindneri* and *L. plantarum* in combination with *S. cerevisiae* or *S. exiguus*, the decrease is about 12.5% of those observed in sourdough. 1-Propanol and diacetyl are not observed, ethanol is almost totally lost, and 2-methylbutanoic acid appears after baking but only when *S. exiguus* is used. The volatile profile is closely related to the yeast species. They contain ethyl acetate and acetaldehyde but differed in alcohol (*S. cerevisiae*) and carbonyl (*S. exiguus*) composition. Isoalcohols and aldehydes increase their relative contribution to the volatile fraction [62]. The volatile fraction of breads with homo- and heterofermentative lactobacilli (*L. plantarum*, *L. delbrueckii*, *L. sanfranciscensis*, and *L. brevis*) and the yeast *C. millen* or *S. cerevisiae* is qualitively similar to that of bread without starters (Table 3), but the former have greater concentration. Principal components of the volatile fraction are acids and alcohols: ethanol, 2,3-methyl-1-butanol, 2-methyl-1-propanol, 1-propanol; 2-methylpropanoic acid, 3-methylbutanoic acid, and acetic acid [137]. Addition of *L. plantarum* to *S. cerevisiae* improves bread flavor, with odor and taste more aromatic and less metallic.

Table 3 Effects of *Lactobacillus* and Yeasts on Volatile Compounds of Bread Crumbs

Bacterial culture		*L. plantarum*			*L. sanfranciscensis*			*L. brevis*		
Yeast added	None	None	*S. cerevisiae*	*C. milleri*	None	*S. cerevisiae*	*C. milleri*	None	*S. cerevisiae*	*C. milleri*
Alcohols										
Ethanol	12010	13000	25038	13813	1965[illegible]	17878	16388	20365	14097	22961
n-propanol	154	82	124	108	75	80	78	107	93	112
2-Methyl-1-propanol	1258	852	1275	1223	1085	1173	1158	1099	1022	1114
n-Butanol	6	n.d.	n.d.	n.d.	n.d.	17	11	3	8	7
2,3-Methy-1-butanol	688	1609	1801	1921	1750	1801	1564	7	7	7
n-Pentanol	7	6	6	6	5	7	n.d.	7	7	7
n-Hexanol	8	9	12	5	n.d.	11	11	12	13	11
Esters										
Ethyl acetate	17	15	11	20	22	20	22	20	18	22
Ethyl	n.d.	3	8	n.d.	5	7	4	4	8	5
Ethyl *n*-hexanoate	3	4	5	7	5	6	7	4	7	n.d.
Carbonyls										
2,3-Butanedione	10	9	n.d.	14	17	20	22	n.d.	11	n.d.
Acids										
Acetic	6	55	37	37	165	344	447	96	122	60
2-Methylpropanoic	34	92	121	81	85	152	119	94	77	65
Butanoic	3	8	7	7	5	10	13	8	7	6
3-Methylbutanoic	29	36	84	52	44	124	63	49	42	36
Others										
Benzaldehyde	7	4	4	5	3	3	3	4	4	n.d.
Furfuryl alcohol	3	5	5	6	n.d.	n.d.	n.d.	n.d.	5	4
Benzyl alcohol	6	4	3	259	11	n.d.	n.d.	8	3	256
2-Phenylethanol	153	210	482	243	15[illegible]	159	235	205	202	214

Data are expressed as relative area of the peak area. 15% sourdough was used.
Source: Ref. 137.

The higher content of these breads in the latter isoacids and some alcohols (2,3-methyl-1-butanol and 2-phenylethanol) can be related to this effect [136]. These results are opposite to those reported before [137] referent to the negative effect of isoacids on bread flavor.

VI. OTHER ASSOCIATIONS (LACTIC ACID BACTERIA AND PROPIONIBACTERIA)

The association of lactic acid bacteria and propionibacteria constitutes an alternative to conventional starters in sourdough processes, providing bread with improved mold-free shelf life [138,139]. *Lactobacillus brevis* and *Propionibacterium* is an association mainly dominated by dough consistency and temperature [138]. The highest propionic acid production (0.55 g/100 g sourdough) is obtained for a mix of *L. brevis* and *Propionibacterium acidipropionici* with a consistency of 320 (g dough/100 g flour), 35 °C, and 32% sourdough addition. However, a properly balanced acid composition requires an additional treatment with α-amylase [140].

VII. FURTHER ASPECTS CONDITIONING INTERACTIONS OF LACTIC ACID BACTERIA AND YEASTS IN SOURDOUGH FERMENTATION

Lactobacilli may possess antimicrobial properties that facilitate their coexistence with other microbial populations, and may account for the predominance of a determined bacteria and for the microbial stability of a sourdough. Bacteriocins (proteins with bacterial activity directed against species usually closely related to the producer bacterium) are common among lactic acid bacteria, and several have been described as produced by different strains of *L. plantarum* [141]. *Lactobacillus sanfranciscensis* also has some antimicrobial properties, which are considered active toward *B. subtilis* but not against sourdough yeast and molds [91]. Two bacteriophages active against *L. fermentum* isolated from wheat sourdough exert an inhibiting effect on some strains of *L. brevis* and *L. fermentum*, reducing and even stopping their fermentative activity [142]. The presence of phages is of interest in continuous processes [66], where the way of reducing the risk of phage infection would be a rotation of bacterial starters composed of strains having different sensitivity toward them.

Other aspects influencing possible associations are growth requirements and resistance to extreme conditions. For example, in spite of its superior characteristics relating to bread quality, *L. sanfranciscensis* is a very fastidious organism with respect to growth requirements and resistance to freezing and freeze drying [50]. Strains of *L. sanfranciscensis* isolated from traditional commercial sourdoughs dye off quickly at extended incubation times, leaving in the starter culture *L. brevis* as a starter organism of inferior quality [50]. It seems that *L. sanfranciscensis* requires a stimulant for growth, identified as a small peptide, containing aspartic acid, cysteine, glutamic acid, glycine, and lysine [143], which may arise in the natural environment of the starter sponge by autolysis of the yeast *C. milleri*. However, more tolerant lactic acid bacteria (*L. brevis, L. plantarum*) will develop and survive in a stressing environment; all of them assure a high and constant number of lactic acid bacteria cells in sourdoughs [144–146] after freezing or freeze drying.

The yeast *S. exiguus* is also a fastidious organism that does not survive after freeze drying; different cultural conditions and/or protective agents do not enhance its survival [129,144]. This restricts its use in preparing commercial microbial starters. Several *S. cerevisiae*, *S. fructuum*, and *C. boidinii* strains show natural tolerance to freezing [144–146], but the fresh cell starter is the only way to enrich the dough with *S. exiguus* [144].

In general, freezing or freeze drying affects the functionality but not viability of starters. A preliminary rehydration and partial reactivation in hydrolyzed wheat flour may be essential for cell functionality. Repeated refreshing and the subsequent use of the initially started sourdoughs enhance fermentation and reduce differences between initially fresh and dehydrated starters [144,147].

REFERENCES

1. G. Spicher and H. Stephan. Handbuch Sauerteig. Biologie. Biochemie, Technologie, 3rd ed. Behr's Verlag, Hamburg, 1987.
2. L. N. Kazanskaya, O. V. Afanasieva, N. A. Komarova, E. A. Yakimycheva, and E. M. Blant, Khlebopek. Konditer. Promst. 8: 21 (1983).
3. L. Kline, and T. F. Sugihara, and L. B. McCready, Bakers' Dig. 44(2): 48 (1970).
4. T. F. Sugihara, L. Kline, and L. B. McCready, Bakers' Dig. 44(2): 51 (1970).
5. M. I. Fields, R. C. Hoseney, and E. Varriano-Martson, Cereal Chem. 59: 23 (1982).
6. R. Calvel, Ind. des Céréales 5: 31 (1981).
7. S. Barber, R. Baguena, M. A. Martínez-Anaya, and M. J. Torner, Rev. Agroquim. Tecnol. Aliment. 23: 552 (1983).
8. S. Barber and R. Báguena, Rev. Agroquim. Tecnol. Aliment. 29/4: 478 (1989).
9. M. Infantes and C. Tourneur, Sciences des Aliments 11: 527 (1991).

10. A. Galli and G. Ottogalli, Annali di Microbiologia ed Enzmologia. 23(1/2/3): 39 (1973).
11. G. Spicher, Brot Backwaren 3: 68 (1989).
12. H. Salovaara and J. Savolainen, Acta Alimentaria Polonica X/3-4: 241 (1984).
13. H. Salovaara and G. Spicher, Getreide. Mehl Brot 41/4: 116 (1987).
14. M. Wlodarczyk, Acta Alimentaria Polonica XI/3: 345 (1985).
15. M. Wlodarczyk, Acta Alimentaria Polonica XII/2: 103 (1986).
16. J. M. Brümmer and K. Lorenz, Cereal Foods World 36: 310 (1991).
17. R. E. Hochstrasser, A. Ehret, O. Geiges, and W. Schmidt-Lorenz, Gebiete Lebensm. Hyg. 84: 581 (1993).
18. R. E. Hochstrasser, A. Ehret, O. Geiges, and W. Schmidt-Lorenz, Gebiete Lebensm. Hyg. 84: 356 (1993).
19. M. Azar, N. Ter-Sarkissian, H. Ghavifeck, I. Ferguson, and H. J. G'harsemi, Food Sci. Technol. 14: 251 (1977).
20. R. A. Tadayon, J. Food Prot. 41: 717 (1978).
21. F. Boraam, M. Faid, J. P. Larpent, and A. Breton, Sciences des Aliments 13: 501 (1993).
22. G. Spicher. Food and Feed Production with Microorganisms, Vol V (H. J. Rehm and G. Reed, eds.). Verlag Chemie, Weinheim. 1983, pp. 1–80.
23. C. Kennes, M. C. Veiga, H. C. Dubourguier, J. P. Touzel, G. Albacnac, H. Naveau, and E. J. Nyns, Appl. Environm. Microbiol. 57: 1046 (1991).
24. M. Gobbetti, A. Corsetti, L. Morelli, and M. Elli, Biotechnol. Lett. 18: 969 (1996).
25. J. M. Brümmer, Cereal Foods World 36: 305 (1991).
26. H. Ng, Appl. Microbiol. 23: 1153 (1972).
27. M. A. Martínez-Anaya, Rev. Esp. Cienc. Tecnol. Aliment. 34: 469 (1994).
28. W. Seibel and J. M. Brümmer, Cereal Foods World 36: 299 (1991).
29. H. Salovaara and T. Valjakka, Int. J. Food Sci. Technol. 22/6, 591 (1987).
30. M. A. Martínez-Anaya, C. Benedito de Barber, and C. Collar, Int. J. Food Microbiol. 22: 249 (1994).
31. P. Weustink, Getreide, Mehl Brot, 43: 49 (1989).
32. M. A. Martínez-Anaya, M. L. Llin, P. Macias, and C. Collar, Z. Lebensm. Unters. Forsch. 199: 186 (1994).
33. K. Lorenz, Bakers' Dig. 55/1: 32 (1981).
34. C. Lynn, Food Technol. 39: 67 (1985).
35. C. Lönner and K. Preve-Akesson, Food Microbiol. 6: 19 (1989).
36. D. E. Van der Werth, F. Lederer, and R. F. Meyer-Roxlau, D. E. Patent 33 35 351 A1. (1983).
37. C. Collar, C. Benedito de Barber, and M. A. Martínez-Anaya, J. Food Sci. 59: 629 (1994).
38. C. Collar, M. A. Martínez-Anaya, and C. Benedito de Barber, Rev. Esp. Cienc. Tecnol. Aliment. 34: 191 (1994).
39. J. M. Brümmer, H. Stephan, and G. Spicher, Getreide. Mehl Brot 38/7: 203 (1984).

40. S. Barber, R. Báguena, C. Benedito de Barber, and M. A. Martínez-Anaya, Z. Lebens. Unters. Forsch. 192: 46 (1991).
41. B. Lund, Å. Hansen, and M. J. Lewis, Lebensm. Wiss. Technol. 22: 150 (1989).
42. C. Lönner and K. Preve-Akesson, Food Microbiol. 5: 43 (1988).
43. J. M. Brümmer and J. Fisher, Ind. des Céréales 66: 27 (1991).
44. W. Röcken, M. Rick, H. Mack, and J. M. Brümmer, Getreide. Mehl Brot 46(5): 139 (1992).
45. W. Röcken, M. Rick, and M. Reinkemeier, Z. Lebensm. Unters. Forsch. 195: 259 (1992).
46. A. Foramitti and A. Mar. Developments in Food Science, Vol. 5B. Elsevier, Amsterdam, 1983, pp. 805–809.
47. A. Galli, L. Franzetti, and M. G. Fortina, Microbiol. Alim. Nutr. 6: 345. (1988).
48. G. Ottogalli, A. Galli, and R. Foschino, Adv. Food Sci. 18(5): 131 (1996).
49. W. P. Hammes, P. Stolz, and M. Gänzle, Adv. Food Sci. 18(5): 176 (1996).
50. W. P. Hammes, Food Biotechnol. 4: 383 (1990).
51. P. Stolz, G. Böcker, W. P. Hammes, and R. F. Vogel, Z. Lebensm. Unters. Forsch. 201: 91 (1995).
52. R. F. Vogel, M. Müller, P. Stolz, and M. Ehrmann, Adv. Food Sci. 18/5: 152 (1996).
53. J. Rossi, Adv. Food Sci. 18/5: 201 (1996).
54. M. J. R. Nout and T. Creemers-Molenaar, Chem. Mikrobiol. Technol. Lebensm. 10: 162 (1987).
55. T. F. Sugihara, L. Kline, and M. W. Miller, Appl. Microbiol. 21, 456 (1971).
56. P. Stolz, R. F. Vogel, and W. P. Hammes, Z. Lebensm. Unters. Forsch. 201: 402 (1995).
57. M. Gobbetti, A. Corsetti, J. Rossi, F-la Rosa, and S-de Vincenzi, Ital. J. Food Sci. 6/1: 85 (1994).
58. M. Iorizzo, R. Coppola, E. Sorrentino, and L. Grazia, Industri Alimentari 34: 1290 (1995).
59. M. Infantes and J. L. Schmidt, Sciences des Aliments 12: 271 (1992).
60. B. Onno and L. Ragot, Ind. des Céréales, 54: 17. (1988).
61. R. F. Vogel, G. Böcker, P. Stolz, M. Ehrmann, D. Fanta, W. Ludwig, B. Pot, K. Kersters, K. H. Schleifer, and W. P. Hammes, Int. J. Systematic Bacteriol. April 223 (1994).
62. M. Gobbetti, M. S. Simonetti, A. Corsetti, F. Santinelli, J. Rossi, and P. Damiani, Food Microbiol. 12: 497 (1995).
63. O. Kandler and N. Weiss, Bergey's Manual of Systematic Bacteriology, Vol. 2 (P. H. A. Sneath, N. S. Mair, and M. E. Sharpe, eds.). Williams and Wilkins, Baltimore, 1986, p. 1209.
64. G. Spicher and C. Lönner, Z. Lebensm. Unters. Forsch. 181: 9 (1985).
65. G. Böcker, P. Stolz, and W. P. Hammes. Getreide Mehl Brot. 49(6): 370 (1995).
66. A. Vollmar and F. Meuser, *Cereal Chem.* 69: 20 (1992).

67. Y. Pomeranz, Wheat: Chemistry and Technology, Vol. 1, 3rd ed. AACC, St. Paul, MN, 1988.
68. P. Savola, H. Salovaara, and J. Enqvist, Developments in Food Science, Vol. 5A. Elsevier, Amsterdam, 1983, pp. 465–470.
69. R. B. Koch, F. Smith, and W. F. Geddes, Cereal Chem. 31: 55 (1954).
70. Y. Pomeranz and K. F. Finney, Bakers' Dig. 39/1: 20 (1975).
71. O. Rouzaud and M. A. Martínez-Anaya, Z. Lebensm. Unters. Forsch. 197: 434 (1993).
72. B. Antuña and M. A. Martínez-Anaya, Int. J. Food Microbiol. 18: 191 (1993).
73. G. Harris and C. C. Thompson, J. Inst. Brewing. 66: 213. (1960).
74. G. Harris and C. C. Thompson, J. Inst. Brewing. 66: 293 (1960).
75. J. Thompson. FEMS Microbiol. Rev. 46: 221 (1987)
76. P. Hautera and T. J. Lövgren, J. Inst. Brewing. 81: 309 (1975).
77. T. Lövgren and P. Hautera, Bakers' Dig. 51(4): 17 (1977).
78. M. R. Siro and T. Lövgren, Bakers' Dig. 53(4): 24 (1979).
79. D. T. Cason, G. C. Reid, and E. M. S. Gatner, J. Inst. Brew. 93(1–2): 23 (1987).
80. J. H. Orlowski and J. P. Barford, Appl. Microbiol. Biotechnol. 25: 249 (1987).
81. M. A. Martínez-Anaya and O. Rouzaud, Food Sci. Technol. Int. 3: 123 (1997).
82. K. K. Shieh, B. J. Donnelly, and B. L. Scallet, Cereal Chem. 50: 169 (1973).
83. R. M. Saunders and H. G. Walker, Cereal Chem. 47: 85 (1969).
84. R. M. Saunders, H. Ng, and L. Kline, Cereal Chem. 49: 86 (1972).
85. M. A. Martínez-Anaya, B. Pitarch, P. Bayarri, and C. Benedito de Barber, Rev. Agroquim. Tecnol. Aliment. 29/1: 63 (1989).
86. E. Giraud, A. Brauman, S. Kelekes, B. Lelong, and M. Raimbault, Appl. Microbiol. Biotechnol. 41: 456 (1991).
87. P. J. Reilly, Starch conversion technology (G. M. A. Van Beynum and J. A. Roels, eds.). Marcel Dekker, New York, 1985, pp. 101–141.
88. A. W. MacGregor, J. E. Morgan, and E. A. MacGregor, Carbohyd. Res. 227: 301 (1992).
89. E. A. MacGregor, A. W. MacGregor, L. J. Macri, and J. E. Morgan, Carbohyd. Res. 257: 249 (1994).
90. H. Neubauer, E. Glaasker, W. P. Hammes, B. Poolman, and W. N. Konings, J. Bacteriol. 176: 3007 (1994).
91. M. Gobbetti, A. Corsetti, and J. Rossi, Adv. Food Sci. 18(5): 167 (1996).
92. B. J. B. Wood and C. Rainbow, J. Biochem. 78: 204 (1961).
93. M. Gobbetti, A. Corsetti, and J. Rossi, Appl. Microbiol. Biotechnol. 42: 939 (1995).
94. M. Gobbetti and A. Corsetti, Z. Lebensm. Unters. Forsch. 203: 82 (1996).
95. O. Kandler, Antonie van Leeuwenhoek 49: 209 (1983).
96. M. Bobillo and V. M. Marshall, Food Microbiol. 8: 153 (1991).
97. F. Götz, B. Sedewitz, and E. F. Elstner, Arch. Microbiol. 125: 209 (1980).
98. M. G. Murphy and S. Condon, Arch. Microbiol. 138: 44 (1984).
99. M. G. Murphy and S. Condon, Arch. Microbiol. 138: 49 (1984).

100. M. Gobbetti, A. Corsetti, and J. Rossi, World J. Microbiol. Biotechnol. 11: 625 (1995).
101. S. Barber, M. J. Torner, M. A. Martínez-Anaya, and C. Benedito de Barber, Z. Lebensm. Unters. Forsch. 189: 6 (1989).
102. M. Gobbetti, A. Corsetti, and J. Rossi, Appl. Microbiol. Biotechnol. 41: 456 (1994).
103. M. Gobbetti, A. Corsetti, and J. Rossi, World J. Microbiol. Biotechnol. 10: 275 (1994).
104. C. Collar, A. F. Mascarós, A. M. Brito, and C. Benedito de Barber, Z. Lebensm. Unters. Forsch. 194: 332 (1991).
105. C. Collar, Food Sci. Technol. Int. 2: 349 (1996).
106. S. E. Gilliland, Bacterial Starter Cultures for Foods. CRC Press, Boca Raton, 1985, p. 42.
107. C. Tourneur, Le lait. 149: 513 (1972).
108. R. C. Lawrence, T. D. Thomas, and B. E. Terzagui, J. Dairy Sci. 43: 141 (1976).
109. M. A. Martínez-Anaya, Adv. Food Sci. 18(5): 185 (1996).
110. C. Collar, A. F. Mascarós, and C. Benedito de Barber, J. Food Sci. 57: 1423 (1992).
111. J. Kratochvil and J. Holas, Getreide Mehl Brot 38/11: 330 (1984).
112. G. Spicher and W. Nierle, Z. Lebensm. Unters. Forsch. 178: 389 (1984).
113. F. Yöndem, M. Özilgen, and T. F. Bozoglu, Lebensm. Wiss. Technol. 25: 162 (1992).
114. G. Spicher and W. Nierle, Z. Lebensm. Unters. Forsch. 179: 36 (1984).
115. A. F. Mascarós, C. S. Martinez, and C. Collar, Rev. Esp. Cienc. Tecnol. Aliment. 34: 623 (1994).
116. C. Collar and M. A. Martínez-Anaya, J. Liq. Chromogr. 17: 3437 (1994).
117. M. Gobbetti, M. S. Simonetti, J. Rossi, L. Cossignani, A. Corsetti, and P. Damiani, J. Food Sci. 59(4): 881 (1994).
118. A. F. Mascarós and C. Collar, Rev. Esp. Cienc. Tecnol. Aliment. 34: 505 (1994).
119. B. K. Dalaly and A. Al-Banna, Iraqi J. Agric. Sci. Zanco 1: 7 (1983).
120. G. Spicher and W. Nierle, Z. Lebensm. Unters. Forsch. 179: 109 (1984).
121. G. Spicher and W. Nierle, Appl. Microbiol. and Biotechnol. 28: 487 (1988).
122. C. Collar and C. S. Martinez, J. Food. Sci. 58: 1324 (1993).
123. C. Collar and C. S. Martinez, Int. J. Food Sci. Technol. 30: 493 (1995).
124. M. A. Martínez-Anaya, B. Pitarch, P. Bayarri, and C. Benedito de Barber, Cereal Chem. 67/1: 85 (1990).
125. Å. Hansen, B. Lund, and M. J. Lewis, Lebensm. Wiss. Technol. 22: 145 (1989).
126. M. Faid, F. Boraam, A. Achbab, and J. P. Larpent, Lebensm. Wiss. Technol. 26: 443 (1993).
127. M. Faid, F. Boraam, I. Zyani, and J. P. Larpent, Z. Lebensm. Unters. Forsch. 198: 287 (1994).
128. J. M. Brümmer, Brot Backwaren. 37(4): 118 (1989).

129. T. F. Sugihara, J. Food Prot. 41: 980 (1978).
130. G. Giovanelli and R. Polo. Ital. J. Food Sci. 1: 71 (1994).
131. P. Damiani, M. Gobbetti, L. Cossignani, A. Corsetti, M. S. Simonetti, and J. Rossi, Lebensm. Wiss. Technol. 29: 63 (1996).
132. M. A. Martínez-Anaya, M. J. Torner, and C. Benedito de Barber, Z. Lebensm. Unters. Forsch. 190: 126 (1990).
133. M. J. Torner, M. A. Martínez-Anaya, B. Antuña, and C. Benedito de Barber, Int. J. Food Microbiol. 15: 145 (1992).
134. Å. Hansen and B. Hansen, J. Cereal Sci. 19: 185 (1994).
135. B. Hansen and Å. Hansen, Z. Lebensm. Unters. Forsch. 198: 202 (1994).
136. Å. Hansen and B. Hansen. Z. Lebensm. Unters. Forsch. 202: 244 (1996).
137. D. Richard-Molard, M. C. Nago, and R. Drapron. Bakers' Dig. 53/3: 34 (1979).
138. P. Javanainen and Y. Y. Linko, J. Cereal Sci. 18: 75 (1993).
139. P. Javanainen and Y. Y. Linko, J. Cereal Sci. 18: 171 (1993).
140. P. Javanainen and Y. Y. Linko, Food Biotechnol. 7/2: 99 (1993).
141. T. R. Klaenhammer, Biochemie 70: 337 (1988).
142. R. Foschino, F. Perrone, and A. Galli, J. Appl. Bacteriol. 79: 677 (1995).
143. R. W. Berg, W. E. Sandine, and A. W. Anderson, Appl. Environ. Microbiol. 45, 786 (1981).
144. L. Cossignani, M. Gobbetti, P. Damiani, A. Corsetti, M. S. Simonetti, and G. Manfredi, Z. Lebensm. Unters. Forsch. 203: 88 (1996).
145. M. A. Martínez-Anaya, B. Pitarch, and C. Benedito de Barber, Z. Lebensm. Unters. Forsch. 196: 360 (1993).
146. M. A. Martínez-Anaya, L. Graña, and M. J. Torner, Rev. Esp. Cienc. Tecnol. Aliment. 33: 399 (1993).
147. M. A. Martínez-Anaya, B. Pitarch, F. J. González, and C. Benedito de Barber, Rev. Esp. Cienc. Tecnol. Aliment. 33: 491 (1993).

5

Baker's Yeast and Sourdough Technologies in the Production of U.S. Bread Products

Karel Kulp
American Institute of Baking, Manhattan, Kansas, U.S.A.

I. INTRODUCTION

The white pan bread, a product standardized by the U.S. Food and Drug Administration, accounted for more than 90% of total wholesale bread production in United States until the middle of the past century. In the 1960s, however, an interest in bread varieties emerged as a result of customers' nutritional concerns and growing demands for diversity of bakery foods. This trend brought about wheat, whole wheat, multigrain, high-fiber, low-calorie, low-salt, and cholesterol-free breads and cereal products [1,2].

The public also demanded premium, more highly flavored cereal foods. This desire was met by artisan breads, introduced by new bakery chains dedicated to manufacturing various types of breads, e.g., hard-crust French, Italian breads and new products including bagels, pretzels, and sweet-dough items (e.g., cinnamon rolls). In the production of the leavened products the application of sourdough fermentation was applied. The new types of bakeries were similar to retail outlets but were organized as franchises and their product selection was more restricted. Often they were combined with a snack shop or a sit-down restaurant operation. The baking operation was on the premises or limited to the bake-off step from frozen doughs, purchased from commercial manufacturers of frozen doughs. The in-store baking in supermarkets became common using frozen doughs. This operation met the demand for fresh bakery products [3,4].

Although the sourdough natural fermentation process originated in prehistoric times leading to three closely related food technologies—baking, brewing, and wine making—its baking application on the North American continent is essentially of European origin. The specifically American process is production of sourdough breads in the San Francisco area that uses microorganisms and procedures different from those used in Europe. This method of bread preparation is described in Chapter 6 and Ref. 5. The sourdough in the United States was applied prior to the introduction of commercial-production baker's yeast as a specific sourdough process or the bread doughs were fermented by yeast from breweries that generally used lactic acid bacteria. Also, long fermentation schedules were conducive to significant activities of lactobacilli present in wheat. Production of rye breads and rye cereal foods also using European methodology, detailed in Chapters 7 and 11, applied sourdough processes.

With the growing preference for better flavor and aroma in breads and cereal foods varieties, the consumer acceptance of sourdough breads in the United States has showed significant growth since the 1980s. According to MRCA, a Stanford, Connecticut, organization that tracks consumer buying and eating habits [6], the consumption of these sourdough breads increased by 98% between 1988 and 1991 alone, reflecting consumers' newly found appreciation for this ancient form of baking. Although additional statistical data are not available, the supermarket bread shelves attest to the growth. Most major bakeries added sourdough bread varieties as well as new sourdough bakery items; many snack shops, fast-food restaurants, and retail baker operations offer excellent sourdough breads and sandwiches; frozen and par-baked products are produced for in-store bake-off. The Internet reports many sandwich recipes using sourdough breads and buns; also a question–answer services offer advice on preparation of sourdoughs as fermentation starters.

This chapter discusses the use and modification of sourdough processes in the production of typical bakery products in United States, covering their microbiological principles, fermentation pathways, technological methodology, and product formulations.

II. MICROBIOLOGICAL PRINCIPLES

Three types of fermentation are active in the sourdough process: alcoholic, lactic, and heterolactic. Naturally occurring yeasts are responsible for the alcoholic fermentation that produces 1 mole of ethanol and 1 mole of carbon dioxide per mole of glucose. The various strains differ in fermentation characteristics due to differences to osmotic pressure sensitivity, acid and temperature tolerances, cell wall permeability, and sugar type fermentability. The yeast strains identified in sourdoughs are summarized in Table 1 [7–9].

Two types of lactic bacteria are active in sourdough processes: homolactic and heterolactic. The homolactic bacteria yield almost exclusively lactic acid as the end product. The heterolactic types yield acetic acid and carbon dioxide in addition to lactic acid. The strains of lactobacilli found in sourdoughs are summarized in Table 2 ([10] and references to individual species: [7, 11], through [20]).

III. METABOLIC PATHWAYS OF SOURDOUGH FERMENTATIONS

A. Alcoholic Fermentation

This fermentation is generated by various yeast strains and proceeds according to the Emden-Meyerhof-Parnas pathway illustrated in Fig. 1 [9]. The basic reaction involve a breakdown of fermentable saccharides present in flour or added to the product formula. Wheat flour contains 1–2% sugars, mainly glucose, fructose, sucrose, maltose, raffinose or melibiose, glucodifructose, and a series of polysaccharides composed of fructose, glucose, and glucofructosans. The major carbohydrate of cereals is starch present as starch granules (about 70% flour, 14% m.b. [moisture basis]) consisting of linear polysaccharide amylose and a branched polymer amylopectin. Both starch components are glucose polymers [21–23].

Of these saccharides, glucose, fructose, and sucrose are fermented at similar rates by baker's yeast that contains active invertase rapidly

Table 1 Yeast Species Identified in Spontaneous Sourdoughs

Species	Ref.
Candida millerii (*Saccharomyces exiguus*, *Torulopsis holmii*)[a]	8,9
Candida crusei	7,9
Pichia saitoi	7,9
Saccharomyces cerevisiae	7,9

[a]Main yeast in San Francisco sourdough bread, utilizing glucose but not maltose.
Source: Data from Refs. 7–9.

Table 2 Identified Species of Lactic Acid Bacteria in Sourdoughs

Species	Homolactic	Heterolactic	Ref.
Lactobacillus plantarum	+		11–14
L. casei	+		11–15
L. buchneri	+		11,13,15
L. acidophilus	+		12,13,15
L. alimentarius	+		16,17
L. farciminis	+		13,14
L. delbrueckii	+		11
L. fermentum		+	11,13,14,15
L. brevis		+	11,13,18,19
L. sanfrancisco (L. brevis var. *lindneri*)		+	13–15
L. pontis		+	20
L. fructivorans		+	7

hydrolyzing sucrose into glucose and fructose already at the dough mixing stage. Yeast strains (e.g., *Torulopsis holmii*) lacking invertase do not ferment sucrose. Maltose is fermented by yeast only at later fermentation stages, generally after the major portion of the hexose monosaccharides has been utilized. The other saccharides of flour are fermented at slow rates or not at all. Fermentation of fructosans was recently reported [24] by Brandt and Hammes demonstrating that yeast invertase is essential to liberate fructose for fermentation by the lactobacilli. Although all yeast strains essentially follow this metabolic pattern, thus producing the same end products, they differ in technological applications due to differences in fermentability of sugars, osmophilic sensitivity, temperature optima for growth and fermentation rates, and variations in fermentation rates. Examples of these properties are baker's yeast, i.e., *Saccharomyces cerevisiae*, the commonly used leavening agent of microbiologically leavened bakery foods, which ferments glucose, fructose, sucrose, and maltose in contrast to *T. holmii* (older name *Saccharomyces exiguus*), which ferments glucose but is unable to utilize maltose. Osmophilic characteristics played an important role in the development of the commercial baker's yeast industry. In prehistorical and historical times, yeast, which is the main supply of carbon dioxide gas in leavened bakery food, was generated by organisms present in the natural cereal grains flora or supplemented from wine or beer fermentations to accelerate the bread process. Since about 1848, brewers in the United States

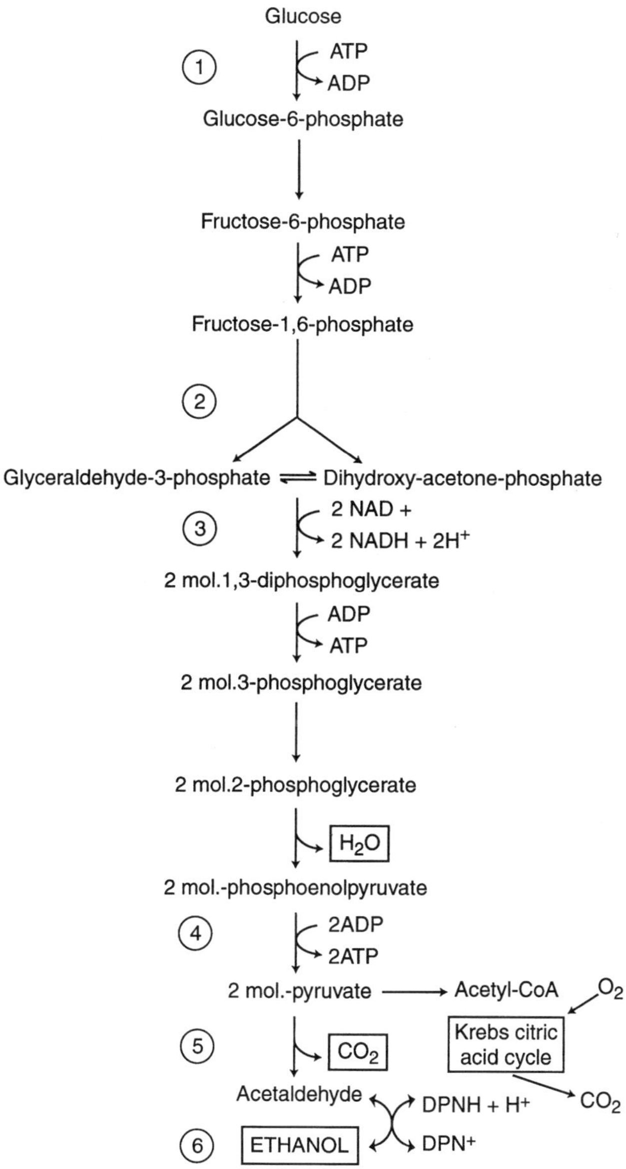

Figure 1 Alcoholic fermentation (glycolysis, Embden-Meyerhof-Parnas pathway) and yeast synthesis. Major enzymes are indicated by numbers: (1) glucokinase, (2) fructose-1,6-diphosphate aldolase, (3) glyceraldehyde-3-phosphate dehydrogenase, (4) pyruvate kinase, (5) acetaldehyde dehydrogenase, and (6) alcohol dehydrogenase. (Modified from Ref. 10.)

started to produce lager beer instead of ale. The lager beer yeast, *Saccharomyces carlsbergensis* (later *S. uvarum*), was not suitable for leavening in bread doughs because it does not tolerate high osmotic pressure. This yeast parameter is also important in selecting yeast strains for fermentation of modern bakery products with varied sugar level in the liquid phase [25,26].

In lean (low sugar) breads, yeasts with low osmophilic tolerance can be used without difficulty. On the other hand, sweet doughs require strains with high osmotic tolerance. In general, the yeast industry selected yeast strains that accommodate the fermentation requirements of the broad spectrum of U.S. bakery food products. Consequently, today's commercial yeasts in the United States show uniform and comparable properties, even when produced by different manufacturers.

The production of baker's yeast exclusively for manufacturing of yeast leavened bakery products dates from later part of the 19th century. It started with production of distiller's yeast using grain mashes as a substrate, replaced later with sugar cane molasses in the United States and sugar beet molasses in Europe as cheaper raw material. This development was made possible by fundamental studies of Louis Pasteur in 1857–1863 proving the microbial origin of fermentation and explanation of its anaerobic nature. The fundamental and technological advances that lead to baker's yeast manufacturing are summarized below [26,27]:

Fundamental advances:

1. In 1680 van Leeuwenhoek observed and sketched yeast.
2. In 1810 Gay-Lussac formulated equation of alcoholic fermentation but failed to recognize its microbiological nature.
3. In 1857–1863 Pasteur proved the microbiological origin of fermentation and explained its anaerobic nature.
4. The years 1850–1900 brought advances of microbiology, such as development of pure cultures and development of propagation of microorganisms by Chrs. Hansen at Carlsberg Laboratories in Denmark.

Technological progress:

1. Use of aeration to stimulate yeast was well known by the end of the 19th century. For example mashes were continuously aerated in Britain in 1886.
2. The fed-batch process (called Zulauf Verfahren) was introduced between 1910 and 1920 by Danish and German scientists. This is the basis of today's yeast production since the gradual feeding of

yeasts during fermentation favors the yield of yeast biomass and minimizes the production of ethanol.

3. In 1890–1892 cereal grains as basic raw material were gradually replaced by cheaper sugar source, e.g., sugar beet molasses [26]. A similar trend took place in the United States where in 1920–1930 sugar cane molasses replaced corn, barley malt, and malt sprouts.
4. Active dry yeast. Progress in drying technology lead to introduction of dry yeast in 1930s. Although it did not become generally accepted in the industry, it found special uses in institutional baking, pizzerias, and home baking.
5. Recent breakthroughs in molecular biology offer new horizons for the yeast industry, permitting production of yeast strains with tailor-made properties, e.g., yeast capable of fermenting lactose, and osmotolerant yeasts for frozen yeasted dough application and/or for use in sweet doughs (more than 20% sugar).
6. During the past two decades there has been a rapid shift to automatic control of fermentation.

IV. PRINCIPLES OF YEAST PRODUCTION

A. Metabolic Pathway During Yeast Reproduction

Alcoholic fermentation is a breakdown of fermentable carbohydrates by yeast under anaerobic conditions according to the Emden-Meyerhof-Parnas mechanism, described earlier and illustrated in Fig. 1 [10]. Under those conditions, yeast generates optimal amounts of ethanol and carbon whereas undergoes only minimal growth as detected by cell multiplication. This reaction also takes place in dough fermentations where only small changes in yeast cell numbers were reported [28].

In yeast production the objective is yeast reproduction, which requires an extensive supply of oxygen in the form of air. The aerobic conditions which are thus created alter the alcoholic fermentation mechanism as depicted in Fig. 1. The glucose breakdown proceeds in both anaerobic and aerobic conditions to pyruvate. In the first case it proceeds to ethanol; in the latter case, the pyruvate via acetyl CoA is coupled to the Krebs citric acid cycle. This metabolic link is also illustrated in Fig. 1.

These reactions differ in energy yields as indicated by following general equations [22]:

1. Alcoholic fermentation:

$$C_6H_{12}O_6 = 2CO_2 + 2C_2H_5OH + 2ATP$$

2. Aerobic fermentation:

$$C_6H_{12}O_6 + 6O_2 = 6CO_2 + 6H_2O + 38ATP$$

B. Production of Commercial Yeasts [7,25,29]

Production of baker's yeast starts with the aseptic transfer of a few thousand yeast cells into a series of progressively larger fermentation vessels ranging from a few gallons to 50,000 gallons or larger in the final stage. A layout of a modern yeast plant is displayed in Fig. 2 (Associated Yeast Plants of Mautner, Markhof and Wolfrum, Inc., Vienna, Austria).

The production of yeast starts in a microbiological laboratory where the suitable strain is selected and then propagated under strictly sterile conditions. Isolated on Petri dishes, yeast cells are grown in sterile media suitable for yeast cultivation. Several transfers are made from small fermentation containers to larger volumes until a sufficient amount of yeast mass is reached for the production-size inoculum.

The production growth medium to which the inoculum is transferred consists of a sterilized blend of clarified sugar cane or beet molasses, minerals, vitamins, and salts. At the initial stage of propagation series all nutrients are added at zero time. The batch stage fermentation in the pure culture tank is still conducted under sterile conditions in enclosed fermentation units without air. Subsequently, the pure culture is than propagated in a yeast starter fermenter.

During the subsequent major propagation steps the nutrients are fed incrementally in synchrony with the growth of the yeast mass. This principle of gradual feeding is called the Zulauf process. The multiplication of yeast cells is favored by intensive aeration, this process uses one volume of air for one fermenter volume per minute of the operation. The schedule of feeding of the nutrients, monitoring and regulation of pH, rate of aeration, and temperature by means of computer programs is designed for specific yeast strain and yeast application. Although establishing absolute sterility conditions of the main process is impossible, a maximal attainable sanitation is essential, involving sterilization of equipment and transfer lines, use of air filters, and general plant maintenance, verified by a well-planned microbiological quality control operation.

The fully fermented mashes contain 4–8% yeast cells, which are recovered by a special type of centrifugal separator (to 50% solids concentration called "yeast cream," chilled to 35–40 °F) and then kept in a refrigerated storage tank until further concentrated by means of plate-and-frame presses or, alternatively, a rotary vacuum filter. Yeast cream may be

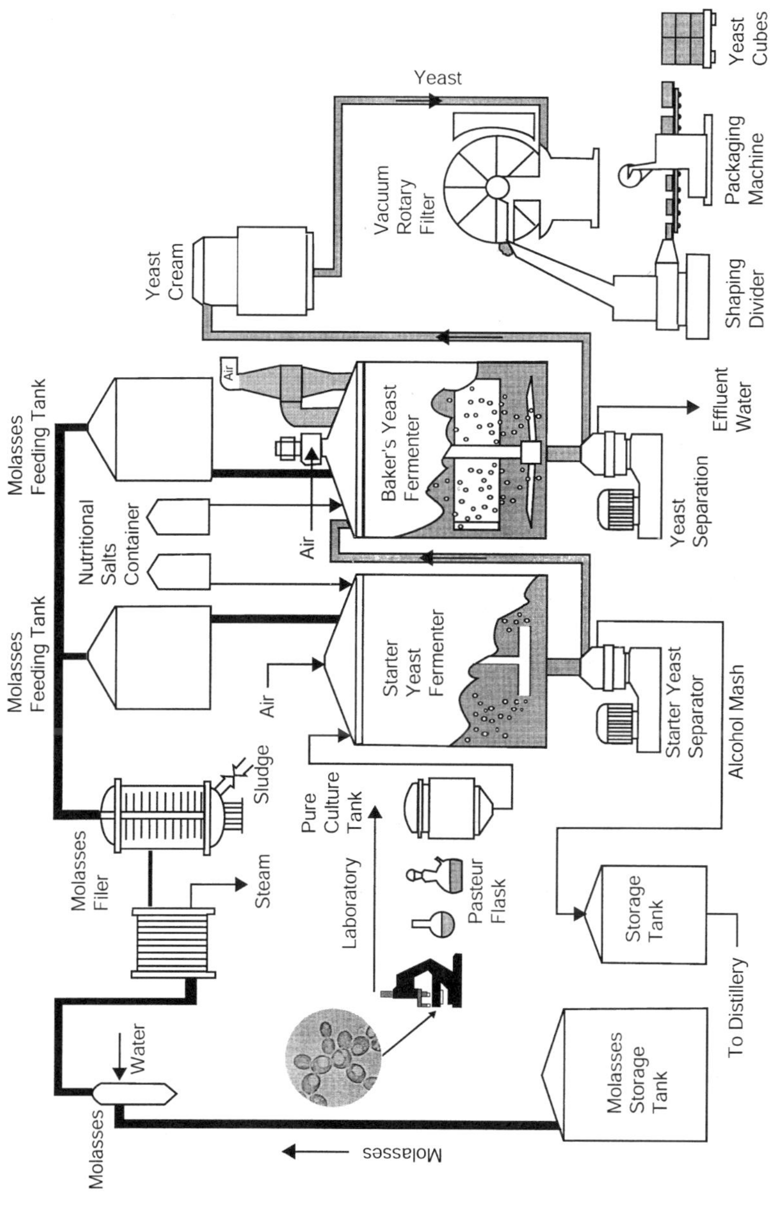

Figure 2 Flow sheet of a yeast plant. From Associated Yeast Plants, Mautner, Markhof and Wolfrum, Ltd., Vienna, Austria, (according to the Komax method).

extensively washed with water in the centrifuges to remove the residues of the growth media and concentrated to a pumpable consistency, refrigerated, and offered to the baking industry in this form for bulk handling. It is generally called "cream yeast" (CmY).

The entire yeast production process takes 5–7 weeks resulting in a yield of thousands of pounds of 30% net yeast solids from the initial few thousand cells. For commercial bakeries, recovered yeast is divided and shaped into elongated 1-lb cubes, or sold in bulk to these bakeries as crumbled cake yeast. Yeasts for home baking are distributed in small packages, generally wrapped in aluminum foil.

Special yeast strains are used for the production of "dry yeast" products, i.e., active dry yeast (ADY), instant active dry yeast (IADY), and protected active dry yeast (PADY). Compressed cake yeast is treated with processing aids or antioxidants prior to drying, Then it is extruded to spaghetti-like threads. After breaking of the threads to 1.5–3.0 cm length, they are dried in a tunnel or conveyor belt type of dryer. This type of drying produces ADY. IADYs with a higher activity than ADYs require gentler drying, which is effected by the fluidized-bed or airlift drying processes. The high activity of IADY is due to high porosity of dried yeast particles, which allows rapid rehydration of yeast in the mixer during dough preparation, eliminating the need for a separate step for rehydration of the dry yeast with water.

PADY contains an emulsifier and is dried by a special process to a moisture level of 5–6.0%. Use of antioxidants (0.1% butylated hydroxyanisole) in this product reduces the adverse effect of oxygen and eliminates the need of special packaging. This type of yeast is useful in complete mixes but requires that the flour moisture level in these mixes be kept low (8–9%); otherwise the yeast loses activity. PADY is also recommended for pizza preparations, generally in a mixture with salt and sugar.

1. Composition and Properties of Yeast Products (Table 3) [29]

As evident from the above discussion and description of production, baker's yeast is marketed in different forms, i.e., compressed yeast (CY), cream yeast (CmY), and instant yeast (IADY). The chemical parameters describing these products are moisture, protein, and phosphorus. Using the strain typified as no. 7752 of the American Type Culture Collection for ADY, it is reported that higher yields of yeast on a sugar basis are achieved, but these are lower in proteins and phosphorus than the CY. However, they generate less carbon dioxide, indicating a lower activity than that in CY which needs to be compensated for by a higher level of yeast solids in order to achieve an equivalent leavening power in bread fermentations.

Table 3 Composition and Shelf Life of Commercial Baker's Yeasts

	Compressed yeast	Cream yeast	Active dry yeast		
			Regular	Protected	Instant
Moisture (%)	70	82	7–8	5–6	4–5.5
Protein, dry basis	60	60	38–48	40–42	39–41.5
Shelf life					
Refrigerated 2–4.5 °C (35–40 °F)	3–4 wk	3–4 wk	6 mo[a] 1 yr[b]	9 mo[a]	1 yr plus[b]
Room temp. 21 °C (70 °F)	Perishable	Perishable	3 mo[a] 1 yr[b]	6 mo[a]	1 yr[a]

[a]In drums or bags, not packaged under vacuum or inert atmosphere.
[b]Packaged under vacuum or inert atmosphere.
Source: Ref. 29.

The carbon dioxide evolution in yeasts is also affected by the level of sugars as evident from the data of Table 4. Higher sugar levels evidently reduce fermentation rates, and yeasts from selected strains are produced for high-sugar–containing sweet doughs [26,29].

Investigations of yeast strains for ADY show that high-protein yeasts can be selected and produced. Although their leavening quality is improved, their stability in drying and storage suffers. Of interest is the presence of trehalose, which should be present in ADY at a minimal level of 12% (d.b. [dry basis]) for adequate stability.

2. Contamination of Commercial Yeasts

As indicated previously commercial yeast is not devoid of microbial contaminants. These microorganisms are not potentially harmful in further yeast applications because many of them are present in natural ingredients, e.g., flours, cereals, and grapes that are fermented by yeast in various technologies. Prior to the production of microbiologically pure yeast, natural starters have been utilized, in certain cases with an advantage as will be further shown in this chapter and other parts of the book. All of the contaminants have to be controlled, and the level of some of them is limited

Table 4 Composition and Activity of Commercial Yeasts (%)

Facts	Compressed yeast	Active dry yeast: Regular	Protected	Instant
Moisture	67–72	7.5–8.3	4.5–6.5	4.5–6.0
Nitrogen[a]	8.0–9.0	6.3–7.3	6.5–7.3	6.3–8.0
P_2O_5[a]	2.5–3.5	1.7–2.5	2.2–2.5	1.8–2.8
Ash[a]	4.0–6.5	4.0–6.5	4.0–6.5	4.0–6.5
Fermentation activity[b]				
In regular doughs[c]	24.5–26.1	15.8–17.4	15.8–17.4	17.9–20.5
In sweet doughs[d]	10.9–12.5	9.2–10.0	9.2–10.0	9.2–10.0
In lean doughs[e]	25.9–38.8	13.6–14.3	13.6–14.3	20.5–21.9

[a]Dry weight basis.
[b]In mM CO_2 produced per gram of yeast solids per hour.
[c]4–12% sugar added.
[d]15–25% sugar added.
[e]No sugar added.
Source: Ref. 25.

by governmental specifications. The U.S. Food and Drug Administration (FDA) requires absence of *Salmonella* and other pathogens, restricting the number of coliform organisms to less than 1000 per gram and the count of *Escherichia coli* to less than 100 per gram. Baker's yeast plants are required to follow Good Manufacturing Practice (GMP). According to Reed and Nagodawithana [7], a predominant portion of contaminants in CY and ADY are lactic acid–producing organisms often called diplos (diplococcci) in the industry. In the past these microorganisms were thought to contribute to the bread flavor. This claim was made by Carlin [30] based on total bacterial counts of $2\text{–}3 \times 10^9$ per gram of yeast but the actual cell count in commercial yeast are actually lower, ranging from 10^4 to 10^8 per gram of yeast. These microorganisms generally belong to heterolactic organisms of the genus *Leuconostoc* or homofermentative bacteria of the genus *Lactobacillus*. In addition to an insufficient level of secondary flora to affect the bread flavor, the short duration of fermentation in U.S. production methods prevents an appreciable formation of flavor compounds in bread processes. Other contaminants that may be identified include some coliform and occasionally a few *Escherichia coli* organisms. Specifications for rope spores (*B. subtilis*) have set an upper limit of 200 per gram of ADY. Other organisms present may be molds, mainly *Oideum lactis*, and contaminating yeasts, such as those belonging to *Candida* and *Torulopsis* genera. Fowell reported methods for detection and estimation of the contaminating microorganisms in yeast [31].

V. PRODUCTION PROCESSES OF WHITE PAN BREAD WITH BAKER'S YEAST

White pan bread remains the major bread product in the United States. As a standardized product, its formulation is subject to certain FDA regulations with regard to composition (e.g., moisture level, vitamin and mineral enrichment standards, use and levels of certain ingredients). The formulation of this type of bread depends on the manufacturing process used; the basic ingredients differ only slightly, except for minor ingredients that are used as dough improvers, crumb softeners, and others. The only leavener used is baker's yeast. Typical formulas of major processes are given in Table 5.

A. Process Characteristics

Sponge and dough is a two-stage process consisting of preparation of a plastic sponge that is fermented with a certain amount of flour, then combined with the balance of flour, water, and other ingredients. *Straight dough* is a single-stage process in which all ingredients are mixed simultaneously and fermented. *Straight dough/No-time* involves dough preparation by chemical dough development by means of reducing agents (e.g., L-cysteine) and minimal fermentation. *Brew (liquid ferment) process* is essentially a sponge-and-dough method whereby plastic sponge is replaced with liquid ferment that can be transferred by pumping. *Continuous* and *Chorleywood* methods are both methods in which the doughs are mechanically developed. The most common process used in the United States is the traditional sponge-and-dough method.

The flow diagram for these processes is given in Fig. 3 [1], describing the duration of the individual processing step for the manufacturing methods.

Table 5 White Pan Bread (%)

Ingredient	Sponge/dough: Sponge	Sponge/dough: Dough	Straight dough	Straight no-time dough	Brew bread
Flour[a]	70.0	30.0	100.0	100.0	100.0
Brew	—	—	—	—	35.0
Water	40.0	24.0	64.0	65.0	32.0
Yeast, compressed	3.0	—	2.5	3.5	—
Salt	—	2.0	2.0	2.0	1.4
Sugar or sweetener solids	—	8.5	8.0	6.0	7.0
Shortening	—	3.0	2.75	2.75	2.75
Yeast food	0.5	—	0.5	0.6	0.5
Nonfat dry milk or milk replacer	—	2.0	2.0	2.0	2.0
Fungal protease	0.5	—	0.25	0.5	—
L-Cysteine, (ppm)	—	—	—	40.0	—
Potassium bromate, (ppm)	—	—	15.0	30.0	30.0
Ascorbic acid (ppm)	—	—	—	60.0	—
Vinegar (100 grain)	—	—	—	0.5	—
Monocalcium phosphate				0.25	
Mono-and diglycerides, hydrate	—	0.5	0.5	0.75	0.5
Dough strengtheners	—	—	—	—	0.2
Calcium propionate	—	0.2	0.2	0.2	0.2

[a]About 11.2–11.7% protein, 0.44–0.46% ash, enriched according to U.S. standards (14% m.b.). All ingredients are given in baker's % (flour = 100%).

Procedures:

1. Sponge and dough: Sponge: Sponge/dough ratio: 70:30; mixing; I min low, 3 min high speeds (77 °F/25 °C); fermentation: 3–5 h (86 °F/30 °C). Dough: Mixing: 1 min low, 10–12 min high (80 °F/27 °C); rest period: 15–20 min. Divide into 18 oz (515 g) pieces for 1 lb (454-g) loaves. Round and give 7 min rest period, sheet, shape, and pan; proof: 55 min. [(107 °F/42 °C)/85% rel. humidity]; bake: 18 min at 445 °F (230 °C).
2. Straight dough: One-step process. Mixing: 1 min at low and 15–20 min at high speeds (78–82 °F/26.5–27.5 °C). Fermentation time 2 hr at 86–95 °F (30–35 °C)/85% relative humidity. This procedure is used by retailers or for specialty breads.
3. Straight dough—no-time: Mix 1 min at low and 10–15 min (80–84 °F/27–29 °C). Proof for 55 min [(107 °F/42 °C)/85% rel. humidity].
4. Brew procedure: Brew: Disperse ingredients by high-speed agitation (5 min, 80 °F/28 °C). Ferment (low agitation) for 1.5 h (88–92 °F/31–33 °C). Add to dough ingredients and proceed as in sponge and dough process. Brew consists of 80.95% water, 7.75% sweetener solids, 8.0% compressed yeast, 3.8% salt, and 0.2–0.5% buffer (calcium carbonate plus ammonium sulfate); 35% of this brew is added to the dough.

Source: Ref. 1.

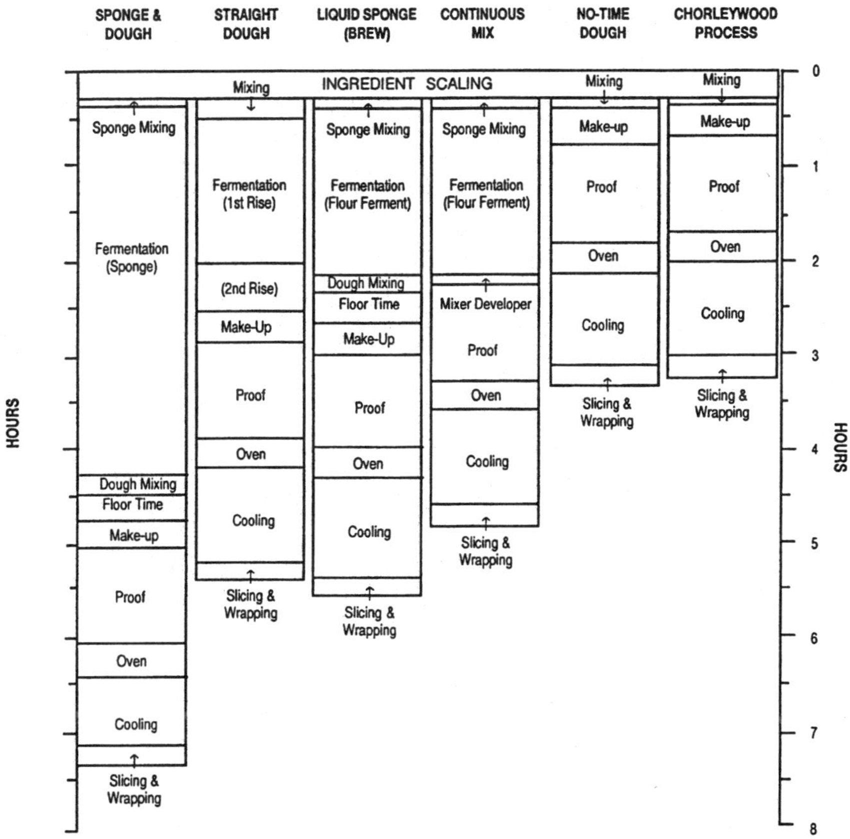

Figure 3 U.S. bread processes with commercial baker's yeast. (From Ref. 1.)

B. Sourdough Fermentations

1. Homolactic Fermentation [10]

Metabolic pathway of homolactic bacteria proceeds by glycolysis (Emden-Meyerhof-Parnas scheme) illustrated in Fig. 4. Exceptions to this mechanism are leuconostocs and group III lactobacilli (these include obligately heterofermentative lactobacilli). Glucose is transported as free sugar and phosphorylated by an ADP-dependent glucokinase to fructose-1,6-diphosphate which is split by fructose-diphosphate aldolase into dihydroxyacetone phosphate and glyceraldehyde-3-phosphate, which is converted to pyruvate through a metabolic sequence including a substrate-level phosphorylation at two sites. Under normal dough conditions, excess sugar, and limited oxygen level, pyruvate is reduced to lactic acid by NAD^+-dependent lactate dehydrogenase (nLDH), thereby reoxidizing the NADH formed during earlier glycolytic steps. Since this pathway produces lactic acid as a virtually sole end product, this fermentation is referred to as homolactic.

2. Heterolactic Fermentation (Fig. 5) [10]

This fermentation has several designations, such as pentose phosphate pathway, pentose phosphoketolase pathway, hexose monophosphate pathway, and, according to *Bergey's Manual* (1986), 6-phosphogluconase pathway. The mechanism involves a dehydrogenation that yields 6-phosphogluconate followed by decarboxylation to form pentose-5-phosphate. The last intermediate is cleaved by the action of phosphoketolase into glyceraldehyde-3-phosphate and acetyl phosphate. Glyceraldehyde-3-phosphate then follows the glycolytic pathway to form lactic acid. In the absence of additional acceptor acetyl phosphate is reduced via acetyl CoA to ethanol and acetaldehyde. Acetic acid may be also formed by this mechanism as indicated in Fig. 5.

Theoretically, homolactic fermentation should yield 2 mol of lactic acid from 1 mol of glucose; the heterolactic pathway is expected to break 1 mol of glucose into 1 mol each of lactic acid, ethanol, and carbon dioxide. In practice, a 0.9 conversion factor is achieved, based on the sugar carbon utilization probably due to the use of sugar carbon in incorporation of cell biomass. Other end products, such as acetic, are also formed as a complex medium such as dough tends to alter the fermentation.

3. Alternative Fates of Pyruvate (Fig. 6) [10]

The key intermediate of the glycolytic and lactic fermentation pathways is pyruvate which is common to all three mechanisms. It also may lead to various other minor intermediate and end products that may contribute significantly to flavor as specific compounds or as flavor precursors.

VI. PREPARATION OF SOURDOUGHS [32–35,39]

Yeast fermentation is essentially a single microorganism action metabolizing sugars. Its main function in bakery processes is leavening by generation of carbon dioxide. Since commercial baker's yeast has been selected and produced to optimally meet this function, it perform this role predictably

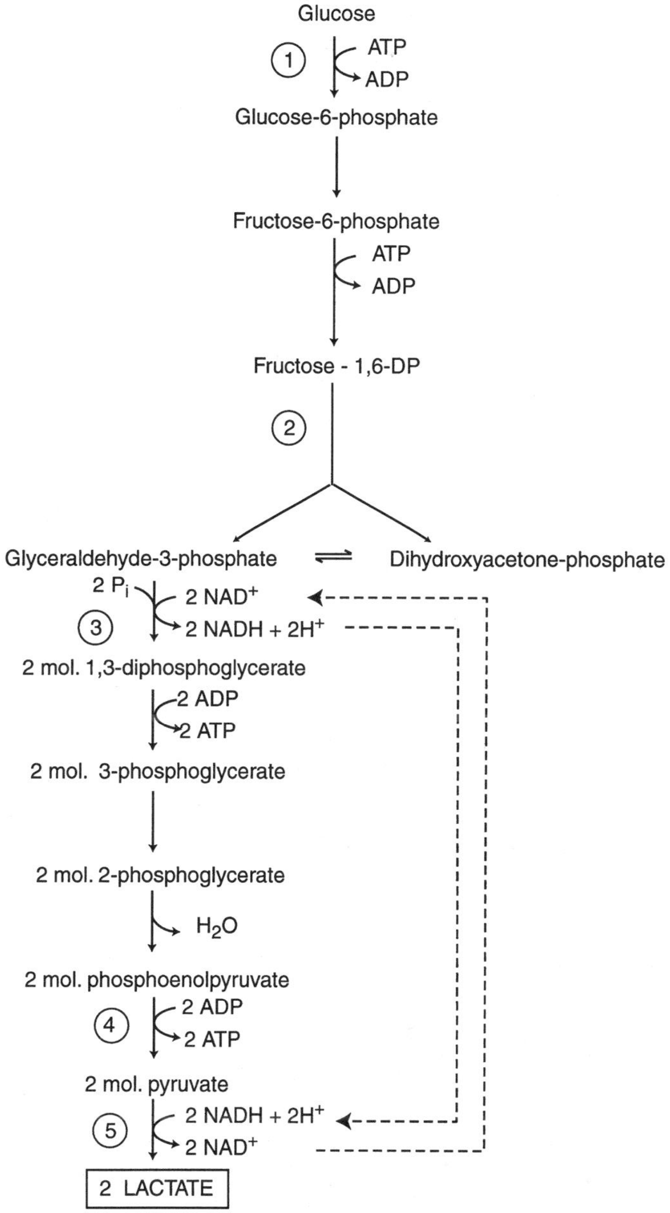

Figure 4 Pathway of homolactic fermentation. Major enzymes indicated by numbers: (1) glucokinase, (2) fructose-1,6-diphosphate aldolase, (3) glyceraldehyde-3-phosphate dehydrogenase, (4) pyruvate kinase, and (5) lactate dehydrogenase. (Modified from Ref. 10.)

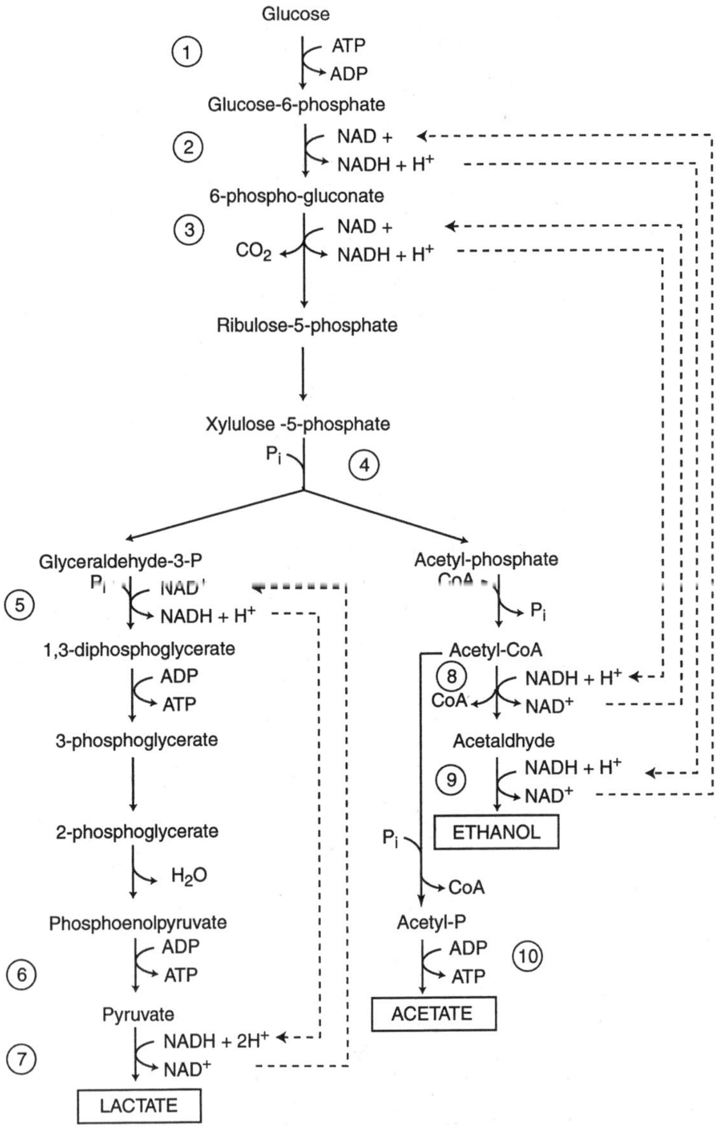

Figure 5 Pathway of heterolactic fermentation. Major enzymes indicated by numbers: (1) glucokinase, (2) glucose-6-phosphate dehydrogenase, (3) 6-phosphogluconase dehydrogenase, (4) phosphoketolase, (5) glyceraldehyde-3-phosphate dehydrogenase, (6) pyruvate kinase, (7) phosphogluconate dehydrogenase, (8) acetaldehyde dehydrogenase, (9) alcohol dehydrogenase, and (10) acetate kinase. (Modified from Ref. 10.)

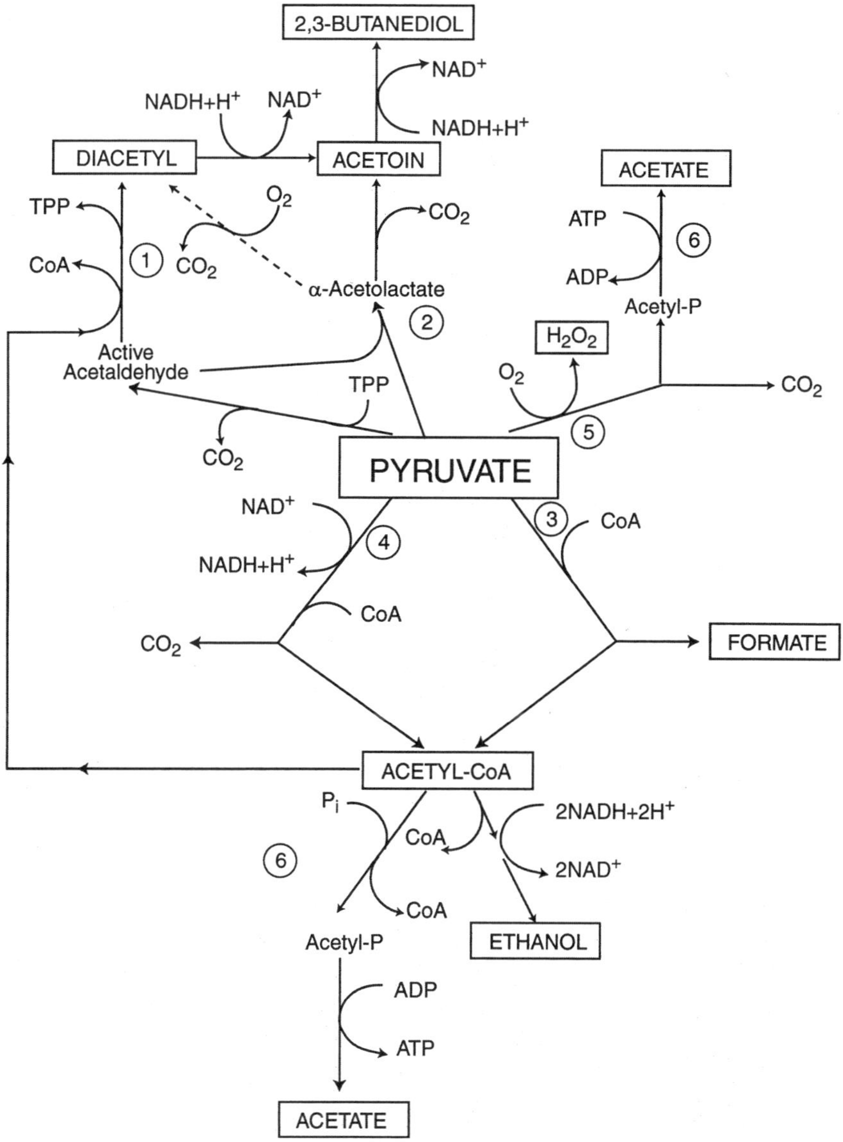

Figure 6 Pathway of the alternative fate of pyruvate. Major enzymes and chemical changes indicated by numbers and arrows (enzymes by numbers and chemical changes by broken arrows): (1) diacetyl synthase. (2) acetolactate synthase, (3) pyruvate-formate lyase, (4) pyruvate dehydrogenase, (5) pyruvate oxidase, and (6) acetate kinase. (Modified from Ref. 10.)

and reproducibly, permitting a rapid wholesale mass production of baker's yeast–leavened products.

On the other hand, sourdough represents a product-mixed fermentation effected by the action of a series of microorganisms naturally present in various cereals. This process is less effective in terms of time but produces breads that are more flavorful and often preferred by some consumers.

A. Methods of Preparation

1. Natural Fermentation

This process involves propagation of the natural flora under optimized media and temperature conditions.

2. Starters Produced by Natural Fermentation

Natural fermentation is used as a processing step in bakeries where the starters are maintained and directly used in production. Sometimes the mature starters, called mother sponge, levain (French), or Anstellgut (German), are refrigerated or frozen and shipped to other bakery plants.

The mature starters are stabilized by drying and sold to the baking industry as ingredients. Depending on the drying procedure, these ingredients may be active or inactive.

3. Synthetic Starters

These products, essentially flavor ingredients, are produced by compounding acids typically present in natural starters.

4. Microbiologically Produced Starters

The starters are generally a mixture of selected lactobacilli grown on various bacteriological media. The lactobacilli are then recovered, washed, and carefully dried to retain their activity.

5. Starting a Sour in Bakeries

Doerry [32,34], who recently investigated baking of breads with a natural sour, described the starting of sour as follows: Only two ingredients—water and flour—are needed to start a spontaneous sour. Although whole grain flour (wheat or rye or its blends) is preferred by most bakers because of the higher microbial count and higher buffering capacity resulting from higher mineral salts content than that of more purified flours, well-functioning starter can be prepared from regular flours. The flours are used when the white color of the product is required. In all cases the microbial flora discussed in Sec. II and in Tables 1 and 2 are adequate for sourdough preparation. Important is the flour/water ratio, which should be kept at 1.25 (water) to 1.00 (flour). The sour of this consistency is similar to a liquid sponge. It can be pumped, cooled, and metered by the same equipment as is used for the liquid fermentation bread process. The development of natural sour is outlined in Table 6, which details flour water ratios, temperature, and maturing times. According to this schedule, the starting flour–water blend is set at 82–85 °F (28–30 °C) and kept at 90–95 °F (32–35 °C) for 24 h. During that time some acidity develops. At that point the flour–water blend must be supplied with additional fresh flour and water. This step is called *replenishment*, a term that will be used throughout the remainder of this chapter. The level of replenishment is computed using a factor called *multiplier*; generally a value of 4 produces good results. The value 4 was used

Table 6 Development of Sourdough Starter

Maturing time (h)	Temp. (°F/°C)	Amount (WU)[a]			
		Starter[b]	Flour	Water	Total
24	95/35	—	1	1.25	2.25
8	95/35	2.25	1	1.25	4.50
16	95/35	4.50	2	2.50	9.00
8	80/27	9.00[c]	16	20	45
16	80/27	45	80	100	225
8	80/27	225	400	500	1125
16	80/27	1125	2000	2500	5625
8	80/27	5625	10000	12500	28125
16	80/27	28125	50000	62500	140625
8	80/27	140625	250000	312500	703125
16–64	39/4	703125	—	—	703125

[a]Weight units (WU) based on any unit: gram (g), kilogram (kg), ounce (oz), or pound (lb).
[b]In first two replenishments, the flour and water are doubled.
[c]In subsequent replenishments, the multiplier 4 is used (the multiplier is the amount of flour and water added to sourdough starter in the replenishment steps). This rate of replenishment increases the total starter fivefold.
Source: Ref. 32.

to calculate the amount of flour and water for replenishment the starter in Table 6. The first replenishments are required after 24 and 32 h maturation, doubling the amount of starter at each interval.

After 32 h when adequate maturity had been reached, the sour can be built up at a higher rate consistently using the multiplier 4, which increases the total amount of sour five times each replenishment step until sufficient sour has been produced. (The computation of replenishment using the multiplier is explained in the note to Table 6.)

Replenished starters are best set and kept at 75–80 °F (24–27 °C). Nonrefrigerated sourdoughs after reaching maturity should be kept at cooler temperatures and must be replenished at least daily. Higher temperatures may favor growth of undesirable microorganisms and lead to development of off-flavors.

Development of a properly matured starter does not only require an achievement of the proper degree of acidity, generally indicated by pH values within 3.6 to 3.8, and total titratable acidity (TTA) of 16–20 mL (TTA in the United States is defined as milliliters of 0.1 NaOH per 20 g of dough, preferment, or mature sour, or 15 g of bread crumb required to raise the pH to 6.6). Even after this stage has been reached, development requires an additional 6–8 replenishments with maturing at 80 °F for full flavor and leavening quality.

B. Current Industrial Methods of Storage [34]

Most large commercial bakeries in the United States are equipped to handle liquid ferments with a flour/water ratio of 1.0:1.5. Mature sours when quickly chilled to 34–36 °F (1–2 °C) are stable and can be used for 10 days without agitation and replenishment during the storage period. Some bakers prefer to store the spontaneous starters at room temperature while replenishing them two to three times per day. The ratio with less water used in these instances produces more viscous starters, which are difficult to pump.

For storage at 39–41 °F (4–5 °C), a 2- to 3-day period is acceptable without subsequent replenishment. If held above these temperatures for a longer time, one or two replenishment steps should be carried out prior to the use of starter for production.

For extended storage, such as several weeks or months, the mature sour can be kept at 39–41 °F (4–5 °C) after "drying up." This process requires a change of the water level to about 30% and that of the flour level to 70% in the sour. This can be achieved by adding flour to the fluid sourdough starter. Before use, the stored starter has to be reconstituted and then replenished a couple of times before being introduced into production.

C. Commercial Natural Starters

Instead of producing the starters in bakeries, commercial starters are manufactured, dried to powders, and distributed to bakeries. The drying process has to be carried out with care to maintain the vitality of the sours. The dry powder of the sours has to be often revitalized by dispersion in water and replenishment prior to its use in production. Optimal use conditions are given by the respective manufacturers.

To illustrate the application of commercial starters several products will be described using the data given in manufacturers' data sheets.

ADM Arkady (Olathe, KA) manufactures several sours containing living souring microorganisms (rye sour, item 112622; and ry-fla-vor, item 112624) and an inactive blend of flavoring ingredients (2 × white sour, item 113514).

Item 112622 is a standardized rye sour culture containing active souring microorganisms, blended with natural flavoring substances (spices, salt, onion powder), recommended for use at levels of 5.5% to 6.5% on the flour weight in "sour dough" (about 1.0% to 1.75% of the total formula flour). The sourdough is set at 80 °F for 22–26 h for revitalization. The rye sour ingredient can be kept in a cool, dry location for 6 months, and for a longer storage below 60 °F.

Rye-fla-vor is highly concentrated unfermented rye flavor, consisting of rye flour, salt, fumaric acid, acetic acid, and spices that does not require a preliminary sourdough preparation. Recommended use levels (flour basis) are 0.5% for American rye, 1.0% for heavy rye (e.g., Pumpernickel, salt rye, and salt rye sticks), and 1.5% for Jewish rye. For wheat, oatmeal, and similar varieties, 0.48–0.78% (flour basis) is advised. Storage stability is 6 months.

The 2 × white sour is similar to the above ingredient. It is a nonfermented product, consisting of corn flour, wheat flour, lactic acid, starch, acetic acid, silicone dioxide, monocalcium phosphate, salt, and dried yeast. The recommended use level is 3–5% (flour basis). Stability is 9 months. In sponge and dough systems it is added on the dough side and in straight dough with flour. It is formulated as a flavor ingredient for breads, rolls, and specialty products.

Bromite Products Inc., Streamwood (Illinois) offers a series of sourdough products produced under controlled bacterial fermentation of doughs. The fermented doughs are dehydrated and ground, and are supplemented with different additives selected for various cereal product groups.

Bro white sour used for San Francisco sourdough bread, bagels, pretzels, English muffins, pizza crusts, bread sticks, snack crackers, pancakes, and so forth is fortified with acidulates, corn syrup, ammonium phosphate, fumaric acid, yeast, ammonium chloride, and natural flavor. This product is highly acidic (pH 3.2 $\pm$ 0.2), granulated to pass USBS No. 20, and it is used at levels from 2.5% to 8.5% of total formula flour depending on the product desired acidity. The sour does not require revitalization. In sponge and dough processes it is added 3–4 min before the end of sponge mixing. For straight doughs this sour is added with the flour.

Bro rye sour is a form of rye sour flavor added to the dough along with other dry ingredients to produce various types of rye breads and buns. This ingredient is a free-flowing natural rye flavor prepared by controlled fermentation of rye flour and yeast. It is dehydrated and ground to be slightly granular (through USDBS #16–100% or USDB#20–90%) and is used at level of 3% or more depending on the desired finished product acidity. Its pH ranges from 4.0 to 4.4. For each pound of bro rye sour, an additional pound of water is recommended.

Bohemian rye sour flavor is a Bromite product similar to that described above, except that it contains buttermilk, has similar pH of 4.20–4.60, and is used at 3–5% for American rye bread, and 8–10% for heavier rye bread (all flour basis). Each pound of sour needs an additional pound of water.

High-acidity pour (pH 2.6 $\pm$ 0.2) is offered for San Francisco–type breads. It is an unfermented composite powder consisting of the following ingredients: wheat starch, flour, citric acid, acetic acid, vegetable shortening (partially hydrogenated soybean and cottonseed oils). Suggested use level is 1.0%, flour basis.

Watson Food company (West Haven, CT) offers three major Diamalt Backmittel (baking ingredients) essentially of European origin. These microbiologically active sourdough dry powders are produced by fermentation of rye flours and provide well-rounded flavors.

DiaRoma powder (F 150015) is a dry rye sour powder that can be used for San Francisco sourdough style breads at 6–8% flour basis; for whole-grain artisan 1–3%, for Italian type 2 to 4%, and for San Francisco style are recommended. These additives are also suitable for straight dough processes.

Lallemand Inc. (Montreal, QC, Canada) with French affiliation produces a series of sourdough ingredients that are marketed in the United States. These are classified as fermented natural wheat sour (LBI 4060) and natural rye sour (LBI 4080). In addition, this manufacturer supplies bakery starter cultures of three type. LBI 4060 is a nontreated fermented wheat sour that has undergone the two-step process and been carefully dried to preserve volatile flavor compounds. This sour contains $<10^4$/g *Lactobacillus*, <100/g *Staphylococcus aureus*, nondetectable levels of *E. coli* and *Salmonella*. The pH of 10% dispersion is 3.2–3.8, acidity 55–70, lactic acid 3.2–3.8%, and acetic acid 0.1–0.3%. It is used in formulas of yeast-raised products, crackers, pretzels, and bagels at levels of 1.0–2.0%.

LBI 4080 is a naturally fermented sourdough that undergoes a three-step process starting from rye whole grain flour and is then dried by a gentle procedure to preserve the volatile components of the sour. The pH of the preparation is 3.2–3.8, acidity 75–85, lactic acid 5.3–6.5%, acetic acid 0.1–0.3%; microbiological specifications indicate the presence of *Lactobacillus* $<10^4$/g, *S. aureus* <100/g, and *E. coli and Salmonella* at nondetectable levels. It is used in production of rye, wheat, wheat-rye breads, yeast-raised goods, crackers, pretzels, and bagels (0.5–1.3% in wheat breads and 0.5–5.0% in rye breads).

Bakery starter culture type I is a blend of lactic acid bacteria (*Pediococcus acidilacti* "MA18/5M") and yeast (*Saccharomyces cerevisiae* "7IB"). The bacterial cell count is $> 2\times10^{10}$/g and that of yeast $>10^9$/g. Listed ingredients include bacteria, yeast, and malt. This starter may be applies in a simple levain or in a two-stage levain process. This process is a traditional method for production of French bread ("pain au levain") using a 24-h sponge (poolish). The inoculation rate varies from 0.2% to 0.03% depending on the levain method chosen.

Bakery starter culture types II and III are similar to type I, except for their bacterial composition. Type II is a blend of *Pediococcus pentosaceus* and the yeast *Torulopsis delbrueckii* "FS", and type III consists of *Lactobacillus plantarum* and *Saccharomyces cerevisiae* "TR." Both can be applied in the same manner and same levels as type I.

VII. FORMULATION OF SOURDOUGH-LEAVENED BAKERY PRODUCTS

A. Ingredients

The essential ingredients of breads and yeast-leavened cereal products are only flour, water and salt. Sourdough modifies these components by fermentation, producing desirable rheological dough structure that expands the dough by gas formation and enhances the product volume. Furthermore, it generates flavor compounds that contribute to the flavor profile or act as precursors during the thermal process in reactions involving the fermentation breakdown products and various flour ingredients. This generally occurs in two types of reactions: Maillard browning (also termed nonenzymatic browning), which involves interactions of fermentation-originated carbonyl compounds with amino groups of amino acids, peptides, and similar compounds. These reactions are aided and accelerated by baking. The baking process produces sugar caramelization, which is thermal decomposition of sugars along with simultaneous formation of polymers. This reaction is the major factor in formation of bread crust and its color. Since the intensity of caramelization and Maillard reaction depend on the pH of the medium (dough) during baking (higher pH favors the reaction and produces higher crust color), the low pH values of sourdough breads necessitate extended baking time to produce adequate crust color.

Salt exerts an inhibitory effect on the microorganisms in the sourdough. As such, its level has to be kept as low as possible, generally within the 1.5–1.8% (flour basis) range. This level is adequate for this type of bread as the salty taste complements the mild acidity.

Minor ingredients used are shortening and vegetable oils, recommended at a 0.76–1.0% flour basis to improve the slicing properties of breads.

Flour type used for starting sourdoughs varies, depending largely on the preferences of baking technologists. Some bakers insists on using wholegrain flour with or without some rye flour to start their spontaneous sour. Generally, a regular flour with an ash content of 0.50% (14.0% moisture basis) suffices for this purpose.

Table 7 shows a typical sourdough bread formula of white pan bread. The amount of starter used in this formula is about 60% (30% of formula flour + 33% of water). The balance of flour (70%) used is a strong spring wheat flour (about 12% protein, 14% m.b.) for high-volume breads. The dough improver used is ascorbic acid, which functions as an oxidant strengthening the dough properties. Addition of small amounts of sugar or pregelatinized starch along with 1% diastatic malt (60° Lintner) enhances fermentation and flavor. Mineral yeast food at addition levels of 0.5–0.75% (flour basis) may improve weak flours. The addition of 0.2–0.8% of compressed yeast (flour basis) or equivalent levels of dry yeast may be required for faster proofing times, but higher yeast levels are not advised because they may produce less desirable bread flavor.

Yeast activity depends on sourdough acidity. Regular baker's yeast is most active within the dough pH range of 4–5. For proof times of less than 2 h, the yeast activity of a dough at pH 3.6 is still satisfactory. A fresh dough prepared according to the formula given above (Table 7) with 30% flour in mature sourdough has a pH about 4.5 and a TTA of 7–9 mL. This pH will drop further to 4.3–4.4 at the divider and the TTA will increase by about 2 mL during the proof time.

The effect of acidity on the dough proof times is shown in Table 8. Rheograph values clearly indicate the diminishing rate of carbon dioxide formation with increasing acidity. As the result of this effect, a reduced rate of dough proofing is expected and observed in actual baking. Vital wheat

Table 7 Sourdough Bread Formula

Ingredient	Formula (% flour basis)
Spring wheat flour	70
Mature starter	60
Salt	1.6
Shortening	1
Water	27
Ascorbic acid	60ppm
Total	162.6

Processing: Mixing: Combine ingredients in the mixing bowl mix until gluten is developed (do not overmix).
Dough temp.: 77–79 °F (25–26 °C).
Bulk fermentation (floor time): 25–30 min (ambient temperature).
Scaling weight: 790 g dough (28 oz) for round loaf.
Final fermentation (proof): 75–100 min at 95 °F (35 °C).
Cutting (slashing): For loaves expected to show much expansion in the oven, cut deeply to prevent wild breaks in the crust Fully proofed loaves are slashed with shallow cuts.
Baking: 35 min at 430 °F (220 °C) with steam at beginning.

Note: Amount added contributes 30% flour +33% water.
Source: Ref. 32.

Table 8 Yeast Activity in Acidified Doughs

	Average rheograph values at time (min indicated)[a]				
Dough pH	30	60	90	120	180
5.64	83	153	227	305	447
4.83	92	168	252	336	463
4.45	95	174	260	346	473
4.00	84	155	226	301	376

[a]Rheograph measures yeast activity by determining carbon dioxide evolution. The higher the number, the higher the activity.
Source: Ref. 34.

gluten is sometimes added to improve the dough strength and to give better chewable crumb properties.

The processing conditions listed in the table differ from those of typical yeast-leavened white pan breads.

Mature sour can be used as a starter (generally in retail operations) or as a sponge (in wholesale production, as illustrated in Table 9).

B. Mature Sour as a Starter

Mature sour is used as a starter and the entire dough becomes a replenishment step. The level of starter added to the dough determines the dough fermentation time: the less starter is added, the longer time required to bring the dough to full maturity and the longer the proof times necessary. Typical amounts of flour in the starter range from 3% to 5%, sometimes as high as 10%, on the total flour formula. The bulk maturation period (floor time) is also affected by the temperature at which the dough is held. However, a temperature below 86 °F favors heterolactic fermentation, producing both lactic and acetic acids; lactic acid generated by homolactic lactobacilli predominates when fermentation is conducted above 86 °F.

C. Maturation in Bulk

This process is equivalent to a straight dough method. The dough is prepared with a very small amount of yeast and kept at ambient temperature. The method produces excellent bread but requires good baking skill to judge the progress and degree of maturation. Production is

Table 9 Replenishing a Mature Sour

	Amount (WU)[a]				Holding	
Replenishing step	Starter	Flour	Water	Total	Temp (°F/°C)	Time (h)
0	—	0.44	0.56	1.0	39/4	—
1	1	1.76	2.24	5.0	80/27	7–9
2	5	8.8	11.2	25	80/27	6–9
	25	—	—	25	39/4	8–160
3	25	44	56	125	80/27	6–9
	125	—	—	125	39/4	8–160
4	Repeat step 3					

[a]Any weight unit, e.g., g, kg, oz, lb.
Source: Ref. 34.

difficult to control and schedule, and is suitable only for small operations. It is also least tolerant with respect to the time/temperature relationship.

D. Mature Sour as a Sponge

This method is well suited for large-scale operations. Table 7 presents the typical formula using 30% of the flour formula in the starter. This level varies with different plants ranging generally between 10% and 40% of the total flour in the dough. For larger bakeries, it is advantageous to use a water/flour ratio of 1.1:1.25. At this consistency, it is possible for the mature sour to be processed, cooled, stored, and conveyed like a regular liquid ferment.

The bulk fermentation time for doughs mixed at 77–80 °F (25–27 °C) usually varies from 10 min [no-time dough (40% or more of the flour is added as part of the mature sour)] to about 2–2.5 h (only 10% of the flour comes from the sour).

French bread and panettone bread formulas including operational variable are given in Table 10.

VIII. OPERATIONAL VARIABLES OF SOURDOUGH

A. Mixing of Sourdough

The acids present in the mature sour increase the gluten dispersion and solubility of gluten proteins and assist with the development of gluten. This results in a significant reduction of dough mixing time and lowers the mixing stability. Consequently, the doughs with added sourdough should be carefully mixed just to the peak consistency and overmixing should be avoided, keeping the dough temperature at 25–26 °C (77–79 °F). The dough is given a floor time (bulk fermentation) for 25–30 min at ambient temperature and then is divided into the desired sizes for proofing.

B. Proofing of Sourdough

It is difficult to develop a procedure with a consistent proof time for bread loaves when the dough is prepared with a small amount of starter (no added yeast) and a relatively long maturation time. Depending on the maturity of the dough at makeup and the proof conditions, proof times may vary from 30 to 90 min. This is acceptable for small retail operations but not for wholesale bakeries. Some bakers even report proof times as long as 4–8 h, which makes production scheduling extremely difficult.

Table 10 Formulas of U.S. Sourdough French Bread and Panettone Bread

	Formula		
	Sourdough French Bread		Panettone bread
Ingredient	20% flour from mature sour[a]	No-time dough[b]	30% flour from mature sour[c]
Bread formula			
Mature sour (1:1.25 water flour)	45	90	67.5
Flour (13.5% protein [14% m.b.])	80	60	70
Salt	1.6	1.6	1.8
Bread shortening	0.8	0.8	1
Ascorbic acid (ppm)	60	60	60
Water (variable)	38	10	27.5
Granulated sugar	—	—	15
Mineral yeast food (bromated)	—	—	0.25
Panettone flavor	—	—	1
Liquid whole eggs	—	—	6
Soaked fruit			
Raisins (midgets)			65
Apricots (diced)			10
Fruit peel mix			5
Water			16
Total:	165.4	162.4	286.05

[a]Processing of French bread, 20% flour from sour: Mix dough to full gluten development (dough temp 78–80 °F (26–27 °C); floor time 1.5–2 h under ambient conditions; proof time 40–60 min at 104 °F (40 °C) and 75–80% relative humidity; baking 35–45 min at 395 °F (200 °C).
[b]Processing of no-time sourdough French bread: same as note a, except that floor time is only 10 min.
[c]Processing of Panettone bread: Soak fruit in advance for 30–60 min; mix dough to full gluten development (dough temp. 77 °F(25 °C); add soaked fruit with all liquids and incorporate by mixing 1 min at low speed and 15 s medium speed; bulk fermentation 3 h at ambient temperature; scaling weight 21.5 oz (610 g); shape dough into lightly round ball (soft dough); put in panettone baking liner; proof about 3 h at 104 °F (40 °C) and bake for 25 min 390–400 °F (200–205 °C).
Source: Ref. 34.

To control proof times within reasonable limits, it is recommended to use the "sponge method" with a large amount of mature starter in the dough. If the baker prefers to use a low level of starter, he may consider adding yeast but should never add yeast to a starter or to a dough that is used as a starter for subsequent sourdoughs.

C. Cutting and Slashing

The fully proofed loaves are cut or "slashed." These cuts are usually about 0.25 in. (6 mm) deep. The first reason for this operation is to prevent the formation of uncontrolled breaks (cracks) in the bread crust during early stages of baking and the second is to improve the appearance of the product.

D. Baking of Sourdough Breads

Most bakery ovens can be used for baking of sourdough breads, if properly adjusted. However, bakers generally prefer ovens with stove hearths because these types of ovens produce the high bottom heat needed for a fast rate of bottom bake for free-standing loaves. At the same time during baking, a low-pressure steam is injected above the loaves to reduce the drying process and thus permit the loaves to expand during the early stages of baking. The steam retains moisture for longer periods in the surface area causing the starch in the crust to gelatinize, thus giving it a shiny appearance ("bloom"). On the other hand, excessive steam depresses the low volume, causing the loaves to spread and producing a tough crust, especially when the oven is not hot enough. The steam level in the ovens can be reduced by temporarily opening the dampers.

When breads are baked on shelves of reel or lap ovens, the shelves may be covered with synthetic stone that produces a bake similar to stone hearth in a deck oven. The loaves can be baked directly on this surface. Alternatively, bakers use trays, frames, or pans manufactured from perforated metal or from a heavy silicone-coated screen-like material. For a better release from flat trays, bakers often line the trays with treated paper or with heavy silicone mat.

Rack ovens are popular because they produce a uniform bake regardless of the loaf position in the oven. However, the crust is somewhat inferior to that of breads baked in deck ovens.

Depending on the type of crust desired, sourdough breads are usually baked for 30–45 min at 210–230 °C (419–450 °F). These temperatures are generally 20 °C (35 °F) less for convection (rack) ovens.

E. Packaging

Crisp crust is a desirable quality attribute in many sourdough breads. To retain this property, these type of breads are often sold in paper bags with open ends rather than in sealed poly bags. Bread in paper bags loses moisture gradually and retains crust crispness, while that in poly bags retains moisture but the crust becomes tough and chewy. The thick-crust breads are generally packaged in paper bags, and the thin-crust breads in poly bags.

Typical white wheat flour sourdough products include San Francisco French sourdough bread (see Chapter 6) which is round bread of 16–20 oz; French bread, called baguettes, which are thin elongated batons about 12 oz; batards, which are a shorter and stubbier version of baguettes; and ficelles, i.e., even thinner and slightly shorter than baguettes.

Generally, many U.S. varieties of cereal foods are produced as specialty bread. These include white pan breads, hamburger buns, bagels, croissants, pretzels, and others. Production of these specialty products is conducted by conventional methods, using different type of sours and technologically modified to reflect the principles discussed here. There are too numerous to be covered in detail, and the reader is referred to special sources, e.g., Ref. 43.

In addition to white wheat flour products, wheat products, multigrain, high-fiber, and other specialty breads and products are fermented or flavored with sours. As an illustration of this type bread, a 60% whole-wheat formula and its processing are given in Table 11.

IX. SPECIAL PRODUCTION LINE FOR SOURDOUGH PRODUCTS

Preparation of sourdoughs in U.S. bakeries is conducted with various types of equipment in vats of different sizes, at room temperature or under controlled temperature conditions. If available, liquid fermentation units, originally designed for liquid fermentation processes, are utilized [36]. A few bakeries use dedicated equipment specially designed for sourdough preparation. They are either custom produced or imported from Europe (mainly Germany, France, and Italy). The equipment is not standardized and is dependent on the processing technology [37,38,44,45].

Physical properties of sourdoughs require certain procedural and equipment modifications. The effect of acids on dough rheology was studied by means of a dynamic sinusoidal oscillation test using a controlled-stress rheometer [40]. These studies confirmed that doughs with acid (no salt) were initially firmer and more viscous than doughs without acids but showed less stability during mixing. These doughs were also more sensitive to overmixing. Addition of salt tended to overcome the adverse effect of overmixing. The commercial mixing operation can be carried out with

Table 11 Sourdough Wheat Bread Dough

Ingredient	60% whole wheat (% flour basis)
Spring wheat flour	10
Whole wheat flour, fine grind	40
Whole wheat flour, coarse grind	20
Mature sour starter	63
Salt	1.8
Liquid honey (optional)	6
Shortening	1
Compressed yeast	0.5
Water (variable)	27
Ascorbic acid	(90ppm)
Total	169.3

Processing:
Mixing: Combine ingredients in mixing bowl and mix until gluten development (do not overmix).
Dough temp. 78–80 °F (26–27 °C).
Bulk fermentation time: 45 min.
Scaling weight: 790 g (28 oz) of dough.
Final fermentation (proof) time: 60–80 min at 96 °F (35 °C).
Cutting (slashing): Fairly deep.
Baking: On oven hearth for about 35–45 min at 420–430 °F (215–220 °C) with steam at the beginning.
Source: Ref. 34.

conventional horizontal mixers provided the mixing operation is carefully adjusted and overmixing avoided. For weaker flours, a spiral-type mixer may be preferred. The mixed doughs are divided either by the conventional ram and shear divider or by a shear procedure. The principles of both methods are discussed by Seiffer [45]. For the ram and shear divider system, the dough is deposited into a dividing machine hopper. A piston pulls back creating a suction that draws the bulk dough into a divider box, readying it for it volumetric dividing. Then, if a knife is present in the system, it is pushed forward, thereby separating the bulk dough from that which is being divided. If no knife is incorporated in the system, process proceeds directly to the dividing stage. In either system, dividing of the dough involves a piston forcing the dough into a sized measuring chamber. The pressure required to force the dough through the chambers is quite abusive and damaging to the dough itself and its cell structure. The action also degasses the dough, modifying the cell structure and consequently yielding a product of a dense, tough, and elastic consistency. Doughs leavened with baker's yeast can tolerate this effect since the yeast readily compensates this loss by rapid fermentation during intermediate proofing. Slow gas formation in sourdoughs causes slow recovery of gas in doughs when this type of divider is used and produces a close and dense crumb cell structure in the finished products. This type of crumb structure is generally less desirable for sourdough and artisan products. The divided doughs receive an intermediate proof, then are rounded, sheeted, molded, panned, given final proofing, and baked.

For the production of sourdough products with typical crumb structure, either the dough is divided manually or a special type of divider using a stress-free dough technology is utilized. Two separate aspects or functions are actually involved in stress-free dividing. The first one is dough dividing. It replaces the conventional extrusion process described above by placing the dough into a processing line in a continuous ribbon with uniform thickness. The second function involves reduction of the dough ribbon to a desired thickness. This is effected by means of a cross-roller, a planetary multibar reduction roll station, and possibly a gauge roller. In general, a typical stress-free processing line combines both aspects. This operation imparts minimal stress on doughs. The only point of stress is when the thickness is reduced and the dough stretched. The ribbon is then cut into specified dough sizes, shaped, and deposited in willow proofing baskets and given final proof. For production of baguettes, a special production line installed in the United States was described by Malovany [46]. In that operation, the divider scales 4- to 24-oz dough pieces in a rectangle, with longer sides of the rectangle extended toward the edges of the conveyor belt. After passing through a belt rounder that stretches the dough pieces to 6 in, they receive 13–20 min intermediate proof.

Subsequent to resting, the dough pieces drop one by one through a presheeter, called "channelize," before they pass through sheeting rollers, a curling chain, and two short pressure plates. The dough pieces are then dropped onto screened pans and proofed. After proofing the dough pieces pass through an automatic scorer and are baked.

A. Bakery Ovens

Bakery ovens and the process were discussed in Sec. VIII. D.

B. Distribution of Sourdough Products

Baked products are distributed freshly baked or partially baked ("parbaked"). Par-baked breads receive about 60% of the full baking time. They are sold in this form or shipped frozen to supermarkets' in-store bakeries for full baking and sales. The baked or par-baked units are directly passed through an equilibrator (vacuum cooler may be used). The freezing step is carried out immediately by blast or cryogenic freezing after the cooling step.

Unbaked sourdough products are generally not offered to the public in retail market. They are shipped to the in-store bakeries for bake-off as frozen doughs. For this application, doughs are produced at the bakeries through the makeup stage, panned, refrigerated, and proofed overnight either in the bakery or during transit in refrigerated trucks. After arrival to the in-store bakery, the doughs are proofed in controlled temperature/humidity cabinets and then baked.

X. QUALITY CHARACTERISTICS OF YEAST-LEAVENED AND SOURDOUGH BREADS

The main differences in bread properties of sourdough and yeast-leavened breads are formula composition, loaf volume, crust and crumb properties, flavor and aroma, and shelf life stability.

A. Formula Composition

As evident from comparison of white pan bread and sourdough formulas (cf. Tables 5 and 7), the latter formula uses simple basic ingredients and avoids various chemical and enzymatic additives present in white bread formulas. Since the sourdough formulation may be formulated with natural ingredients only, including starters derived from indigenous microbial flora, this process is suitable for the development of natural bakery products.

B. External Bread Characteristics

White pan bread is characterized by large volume and relatively thin crust. The volume is attributed to the use of strong flours, the presence of high levels of fermentable sugars, and the vigorous rate of fermentation by baker's yeast. The thin crust is due to the dough expansion during the baking process and the use of shortening and surfactants in the formula.

Sourdough breads are characterized by lower volumes and thick and tasty crusts produced during longer baking times and application of steam.

C. Internal Bread Characteristics

The crumb grain of conventional baker's yeast–fermented breads is uniform, and its cell walls are thin due to the uniform gas distribution in the doughs throughout the thin gluten film induced by the dough makeup procedure. The yeast fermentation is vigorous and occurs during the entire process within the optimal yeast fermentation range from pH 4.8 to 5.2. Another important attribute of the crumb is its texture and softness, which is associated with freshness.

On the other hand, loaf volumes of sourdough breads are smaller, their crumb grain is open and nonuniform, giving the breads a natural rustic appearance.

D. Keeping Properties of Breads

Baked products are perishable. They undergo physicochemical changes, generally referred as "staling" and microbial degradation. The term *staling* designates the loss of consumer acceptability due the storage changes other than those caused by the microbiological action. The main change that reduces the bread quality during aging is crumb firming, which is attributed to starch retrogradation. This process is generally retarded in conventional breads by addition of certain surfactants and enzymes. The mechanism of the bread staling has been studied extensively, and present views are summarized in recent publications [41,42]. In a simplified manner, starch gelatinizes during baking and the amylose component leaches out from the granules. Upon cooling the amylose crystallizes and determines the firmness of fresh bread crumbs; amylopectin crystallizes (retrogrades) at a slower rate than amylose and causes gradual firming during storage. This change is highly correlated with the consumers' perceived freshness as determined organoleptically. Other factors that affect freshness judgment are a the loss of moisture due to simple drying out, which can be controlled by moisture-proof wrapping, or moisture migration from the high moisture crumb phase to the dry crust. The latter change is minimized by production of thin-crust breads. Two types of reactions affect flavor deterioration associated with aging of breads: (1) transfer of certain flavor components from the crust to the crumb region and (2) loss of some flavor compounds by formation of complexes with amylose [42]. The crust of wrapped products loses its fresh crispness due to equalization of moisture.

Sourdough breads stale more quickly than conventional breads. Having generally smaller specific volume, sourdough crumb—even of freshly baked breads—is firmer. Although no objective studies on the staling rates of this type of bread are available, it can be postulated that the firming rates are higher due to the absence of antifirming ingredients in the formula, with reduced enzymatic activity of α-amylases due to low pH conditions in doughs. Furthermore, the low-acidity conditions and longer baking times are likely to affect the starch properties by acid hydrolytic modification of the starch flour component that generates higher level of soluble starches and modifies the retrogradation properties of the swollen granules. The thick crusts of various sourdough breads also contribute to the aging process by dehydrating the crumb by moisture equalization. To maintain the bread crumb crispness (a desirable attribute of this type of bread), the sourdough breads are sold both unwrapped and wrapped in special perforated films. In the United States, the crisp crumb is not an essential quality component. Consequently, wholesale breads, even baguettes, are sold wrapped with soft crumbs.

In contrast to quick firming, sourdough breads exhibit good resistance to microbiological deterioration. This extended shelf life is attributed to the presence of acetic acid. The conventional breads that are susceptible to microbial damage require additions of antimicrobial agents, e.g., calcium propionate, to prevent microbial growth.

XI. FLAVOR OF YEAST- AND SOURDOUGH-LEAVENED BREADS [47–48]

The flavor of both types of bread consists of the odor of freshly baked breads and of flavor components retained in the crust and crumb and perceived organoleptically upon tasting. The odor of breads dissipates throughout the bakeries during the baking process and is considered to be highly desirable, being readily recognized and associated with bread baking. The odor from sourdough bread baking is essentially similar but more pungent due to the presence of acids in the volatiles. The acidic acid is the major acid in this fraction.

Bread flavor in either bread type cannot be attributed to a single compound. This was recognized by various investigators, and a long list of flavor components was generated by identifying them by various analytical methods. The compounds reported or produced during fermentation (baker's yeast) and/or baking are listed in Table 12 [50]. These include organic acids, carbonyl compounds, products of Maillard reaction or thermal degradation, and interaction of fermentation intermediates with amino acids. In spite of these studies, no bread flavor blend was successfully prepared from these compounds. An interesting observation was made by Wiseblatt et al. [51] who obtained bread-like aroma by heating dihydroxyacetone with proline. This reaction was further studied by Hunter et al. [52]

Table 12 Flavor Compounds Generated by Yeast Fermentation and Baking

Organic acids		Alcohols	Carbonyls	Products of Maillard reaction
Butyric	Acetic	Ethanol	Acetaldehyde	Furfural
Succinic	Lactic	*n*-Propanol	Formaldehyde	Hydroxymethylfurfural
Propionic	Formic	Isobutanol	Isovaleraldehyde	2-Methylpropional
n-Butyric	Valeric	Amyl alcohol	*n*-Valeraldehyde	3-Methylpropional
Isobutyric	Caproic	Isoamyl alcohol	2-Methyl butanal	3-Methyl butanal
Isovaleric	Caprylic	2.3-Butanediol	*n*-Hexaldehyde	Methional
Heptanoic	Isocaproic	2-Phenylethyl alcohol	Acetone	dihydroxyacetine
Pelargonic	Capric		Propionaldehyde	
Pyruvic	Lauric		Isobutylaldehyde	
Palmitic	Myristic		Methyl ethyl ketone	
Crotonic	Hydrocinnamic		3-Butanone	
Itaconic	Benzylic		Diacetyl	
Levulinic			Acetoin	
			Methional	
			Glyoxal	
			3-Methyl butanal	

Source: Ref. 50.

who confirmed this investigation and prepared similar flavors by heating of proline with glycerol. These investigators identified the reaction product of this reaction as a derivative of pyrrolidine (*N*-methyl-2-acetopyrrolidine) and postulated Wiseblatt's to be similar to their product. They also detected these flavors in bread extracts by means of a specially modified gas chromatography observing three aroma peaks [52].

Sourdough bread flavor was also studied but no systematic data are available for comparison of the flavor composition of yeast- and sourdough-fermented breads, partly due to differences in analytical and sample preparation methodology but mainly because sourdough flavor investigations focused on the study of the doughs rather than on baked breads as in the case of yeast-fermented breads that were analyzed after baking. Consequently, the missing effect of baking in the sourdough studies on the modification of flavor compounds and generation of heat-formed components prevents a proper comparison of flavor composition between these bread types.

Frasse et al. [53] analyzed the volatile compounds in French sourdough breads. Three types of dough were investigated by gas chromatography–mass spectrometry and by gas chromatography–olphactometry for composition of aromatic compounds in corresponding extracts. These doughs included control dough (YF) prepared by a straight dough process and fermented by baker's yeast (*S. cerevisiae*), yeasted but not fermented dough that was analyzed immediately after mixing (YNF), and nonyeasted but fermented dough (NYF; treated in the same manner as the control dough YF but without addition of yeast in the formula). This experimental design permitted evaluation of the contribution of yeast and indigenous flour microorganisms by fermentations to dough flavor. In total more than 40 volatile compounds were detected, among which 20 were alcohols, 7 esters, 6 lactones, 6 aldehydes, 3 alkanes, and 1 sulfur compound. Except for the aldehydes and the alkanes, all of these classes of compounds increased with action of *S. cerevisiae*, especially the alcohols that are referred to as fusel alcohols. The detected compounds we screened for flavor contribution by gas chromatography–olphactometry. These results are shown in Table 13. The 2,3-butanedione (diacetyl), 3-methyl-1-butanol, 2-methyl-1-butanol, methional, 2-phenylethanol, and two unindentified

Table 13 Most Pronounced Flavor Compounds in NYF (Nonyeasted but Fermented), YNF (Yeasted but Nonfermented), and YF (Control Yeasted and Fermented) Doughs

		FD factor[a]		
Compound	Odor description	NYF	YNF	YF
Unknown 13	Pungent	—	512	1024
3-Methyl butanal	Malty	2	256	256
2,3 Butanediol (diacetyl)	Buttery	2	256	1024
Hexanal	Green, grassy	16	2	2
Unknown 14	Rubber-like	—	2	16
3-Methyl butylacetate	Banana-like	—	—	32
3-Methyl 1-butanol + 2-methyl 1-butanol	Malty	2	64	128
Unknown 15	Musty, pungent	32	6	4
Unknown 16	Ferment, straw	—	—	64
Unknown 17	Mushroom-like	32	32	128
Trimethylpyrazine	Potato-like	2	16	64
Methional	Cooked potato	128	256	512
Furfural	Sweet, bread	4	128	32
Unknown 18	Musty, roasted	2	32	32
Unknown 19	Green, pungent	16	4	4
trans-2-Nonenal	Green, tallowy	256	512	512
γ-Caprolactone	Green, pungent	128	8	8
trans, trans-2,4-Decadienal	Fatty, waxy	2048	1024	—
2-Phenyl ethyl alcohol	Wilted rose	2	128	512
γ-Nonalactone	Coconut-like	4	4	16

[a]Flavor dilution obtained by aroma dilution analysis (AEDA).
Source: Ref. 53.

compounds with a pungent and mushroom-like odor were the most pronounced odorous compounds generated during fermentation of the dough, as shown by the aroma extract dilution analysis of the three aromatic extracts.

A similar study conducted by Czerny and Schieberle was recently reported [54]. This investigation confirmed that sourdoughs fermented with whole-wheat flour are richer in flavor volatiles than those with flour type 550 (flour milled to ash content 0.55% according to German standards), as evident from Table 14. These investigators further succeeded in correlating certain compounds with the fermented sourdough aroma. Accordingly, the stronger fatty aroma of whole-wheat dough coincides with the presence with fatty flavor of isolated compounds in Table 14. Furthermore, a correlation between the musty odor character of 2-/3-methylbutyric acid as well as the malty flavor of 3-methyl butanal were detected. The difference between the tested flour types is most likely due to the presence of different levels of certain flavor components already occurring in the flour and to the metabolic changes that take place during lactic fermentation. The recognition of these flavor components in flours may be beneficial in the selection of flours for breads of improved flavor quality.

Table 14 Comparison of the Intensities of Important Flavor Compounds of Sourdoughs from Whole Wheat Flour and from Flour Type 550 which Differ in Minimally Four dilution Levels

		Dilution factor	
Flavor compound	Aroma	Whole wheat	Flour 550 dough
trans-2,4-Nonadienal	Fatty, roasted	512	4
Phenylacetaldehyde	Bees wax	256	4
Methional	Cooked potatoes	1024	16
trans,trans-2,4-Decadienal	Fatty, roasted	512	16
4-Vinylmethoxyphenol	Lilacy	512	16
3-Hydroxy-4,5-dimethyl-2(5H) furanon	Sweaty	4096	256
trans-4,5-Epoxy-*trans*-decenal	Metallic	4096	256
trans,cis-2,6-Nonadienol	Cucumber, fatty	1024	64
trans,cis-2,6-Nonadienal	Cucumber, fatty	1024	64
Vanillin	Vanilla	1024	64
2-/3-Methylbutyric acid	Sweaty	256	16

Source: Ref. 54.

Important variables affecting the production of flavor compounds, i.e., wheat flour type and sourdough microbial flora, were evaluated by Hansen and Hansen [55]. Three wheat flour grades were tested: straight grade (14.3% protein, 0.66% ash), low-grade (16.8% protein, 1.80% ash), and whole-meal (13.6% protein, 1.64% ash), and four starter cultures of *Lactobacillus* spp. were investigated.

The volatile compounds and acids are formed in higher amounts by sourdough fermentation of low-grade flours than of the refined flour. However, the same level of acids and the same content of volatile compounds are produced in dough from low-grade and whole-meal flours. Flour type is the main factor in acidification of sourdoughs, and the production of lactic acid is highest in sourdoughs made from low-grade and whole-meal flours. On the other hand, the starter culture is the main factor in the production of volatile compounds, including acetic acid.

The flour type has a significant effect on the formation of ethyl acetate and ethanol in sourdough fermented with heterofermentative cultures, especially *L. sanfrancisco*, with highest amounts detected in whole meal and low-grade flour sourdoughs. The production of alcohols other than ethanol was greatest in sourdoughs from straight-grade flour. Apart from ethyl acetate, only a few esters were formed in small amounts. Carbonyls were detected only in the homofermentive culture.

XII. CHEMICAL AND BIOCHEMICAL REACTIONS GENERATING FLAVOR OF BREADS

From the above data of flavor compounds it can be concluded that the following types of reactions contribute to flavor formation.

A. Alcoholic and Lactic Fermentations

As evident from pathways illustrated in Figs. 1, 4, and 5, the fermentable carbohydrates produce a series of intermediate compounds which can serve directly as flavor components of the bread or dough flavor complex or can act as precursors of flavor compounds generated in the baking process.

B. Enzymatic Reactions

In addition to the metabolically induced changes referred to in Sec. A above, enzymes of yeast, the flour microflora, and the flour itself produce compounds that interact in the baking process and form flavor components. Cereal amylases yield mono- and oligosaccharides from milled damage starch and participate in caramelization and Maillard reactions along with reducing sugars.

Caramelization is a reaction that involves reducing sugars (mono-, di-, and oligosaccharides) only and takes place when in the temperature range 120–240 °C and with a humidity content of 5–10% in the crust bread region. All phases of the mechanism of the Maillard reaction have not been elucidated. In the first stage, the reaction proceeds by the interaction of the free amino group of amino acids and the carbonyl group of monosaccharides. Additional complicated steps lead to production of furfural (from hexose sugars) hydroxymethylfurfural (from pentoses), and amino acids. The amino acids can be further transformed into aldehydes of one carbon less than the respective amino acid. Obviously, this reaction can be a source of flavor compounds.

C. **Formation of Higher Alcohols** (Fig. 7) [56,57]

Originally the formation of higher alcohols was attributed to enzymatic transformation of amino acids by Ehrlich [56]. His proposed mechanism involved several enzymes, such as transaminase, which transfers the amino group of the amino acid into an acid and itself forms a ketoacid. The ketoacid is then transformed into an aldehyde by the action of carboxylase through the loss of carbon dioxide. Finally, the action of the enzyme alcohol dehydrogenase reduces the aldehyde to alcohol. This reaction is illustrated in Fig. 7A with transformation of L-leucine into isoamyl alcohol.

This mechanism was insufficient to explain formations of all alcohols to be responsible for the entire levels of some of them. An alternate mechanism was revealed by Ayrapaa [57]. According to his suggested pathway, the ketoacids produced by carbohydrate catabolism are intermediates in the formation of higher alcohols. In this pathway, isoamyl alcohol can be generated from pyruvic acid or 2-ketoisovaleric acid, both of which are intermediates of carbohydrate metabolism. This process is illustrated in Fig. 7B [57].

XIII. CONCLUDING REMARKS

This chapter has included the following:

1. We discussed the present market of breads and cereal products and the application of sourdough technology. We observed the growing preference for variety products, including those produced by sourdough fermentation.

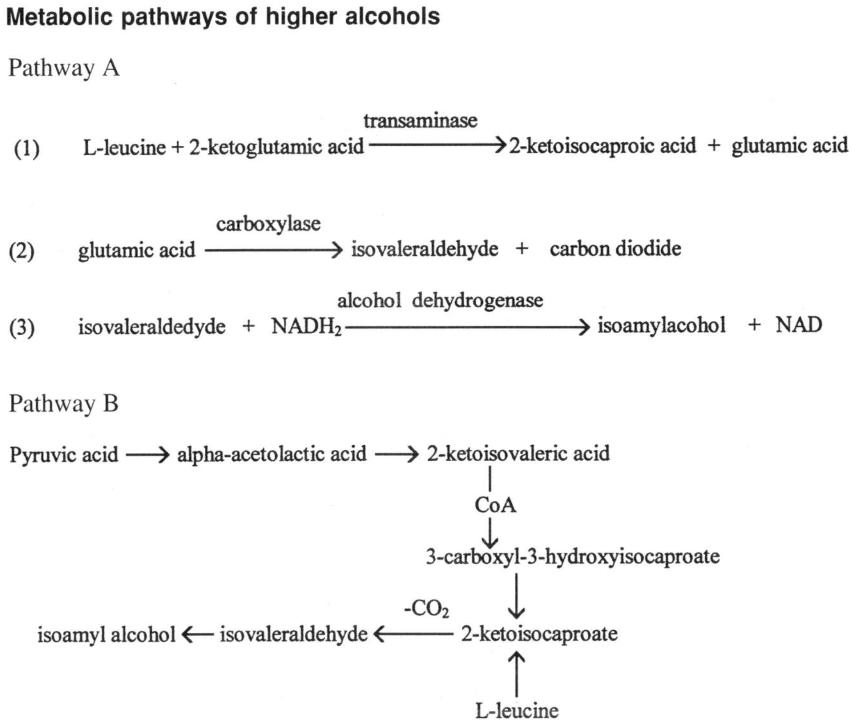

Figure 7 Pathway of formation of higher alcohols. (A) From amino acids by transamination [56]. (B) Via pyruvate [57].

2. The commercial baker's yeast fermentation remains the predominant fermentation process due to its reliability and predictability in mass production of breads. It was presented by giving salient historical developments, description of commercial yeast manufacturing, its uses and properties, and presentation of main bread manufacturing methods used in the United States. The fermentation pathways of carbohydrate metabolism and yeast multiplication were also described.
3. Sourdough technology was discussed with respect to the microbial flora and lactic acid fermentation, including the metabolic pathways of homolactic and heterolactic lactobacilli. Processes of preparation of natural and commercial starters were discussed and formulation of breads utilizing both types of starters presented.

4. Processing variables and rheological effects of lactic acid fermentations were presented and related to dough handling and equipment requirements.
5. Finally, the characteristics of yeast-leavened bread and lactic acid–fermented breads were described, covering internal and external loaf quality parameters, staling rates, microbial storage stability, and flavor. The last property was covered in some detail, with emphasis on our present knowledge of bread flavor formation by fermentation and thermal reactions.

REFERENCES

1. Kulp, K. and Ponte, J. G. Breads and yeast-leavened bakery foods. In: Handbook of Cereal Chemistry and Technology (K. Kulp and J. G. Ponte, eds.). Marcel Dekker, New York, 2000, pp. 539–565.
2. Anon. Artisan-style naturally-leavened bread making a market impact. Milling and Baking News, January 26, 1999, 30–32.
3. Lorenz, K., and Kulp, K. Freezing of dough for production of breads and rolls in the United States In: Frozen and Refrigerated Dough and Batters (K. Kulp, K. Lorenz, and J. Bruemmer, eds.). American Association of Cereal Chemists, St. Paul, MN, 1995, pp. 136–153.
4. Kohn, S. An update of the U.S. baking industry. Cereal Foods World 45:94–97 (2000).
5. Hoerner, G., and Boge, T. San Francisco style sourdough bread. Am. Inst. Baking Res. Bull. (19); 1997.
6. Kroskey Meres, C. The sour smell of success. Bakery Production and Marketing, November, 26, 1994, 24–30.
7. Reed, G., and Nagodawithana, T. W. Yeast Technology, 2nd ed. Van Nostrand Reinhold, New York 1991, pp. 315–368.
8. Kline, L., and Sugihara, T.F. Microorganisms of the San Francisco sour dough process. II. Isolation and characterization of undescribed bacterial species responsible for the souring activity. Appl. Microbiol. 21:459–465 (1971).
9. Salovaara, H. Functions of microorganisms in sour doughs: an overview of recent research, in Proc. Int. Symp. Cereal Technology in Sweden, Ystad, Sweden, 1988, pp. 91–101.
10. Salovaara, H. Lactic acid bacteria in cereal-based products. In: Lactic Acid Bacteria (S. Salminen and A. von Wright, eds.). Marcel Dekker, New York, 1993, pp. 111–126.
11. Kazanskaja, L. N., Afanasyeva, O. V. and Pali, V. A. Microflora of rye sours and some specific features of its accumulation in bread baking plants of USSR. In: Proc. World Cereal Bread Congress, Cereal 82nd Symp., Prague, Development of Food Sciences, 5B, 1983, pp. 759–763.

12. Salovaara, H., and Kalumpaa, H., An approach to the classification of lactobacilli. Isolation from Finnish sour rye ferments, Acta Alim. Polon. 10:231–239 (1984).
13. Spicher, G. Microflora of sour doughs. XVII. Communication. Further investigations of the variability of commercial sourdough starters. (German), Z. Lebensm. Unters. Forsch.: 106–109 (1984).
14. Spicher, G., and Loenner, C. Microflora of sour doughs. XXI. Communication. The lactobacilli occurring in sourdoughs of Swedish bakeries. (German), Z. Lebensm. Unters. Forsch. 181:9–13 (1985).
15. Foramitti, A., and Mar, A. New aspects and observations in continuous production of sourdoughs in the bread industry. Getreide Mehl Brot 36:147–149 (1982).
16. Loenner, C., Welander, T., Molin, N., Dostalek, M., and Blickstad E. The microflora in a sour dough started spontaneously on typical Swedish rye meal. Food Microbiol. 3:3–12 (1986).
17. Spicher, G., and Schroeder, R. Microflora of sour doughs. IV. Communication. Investigation of the method of "pure culture sour" of the occurring rod-forming lactic bacteria (genus Lactobacillu Beijerinck) (German). Z. Lebensm. Unters. Forsch. 167:342–354 (1978).
18. Azar, M., Ter-Sarkissian, N., Ghavifek, H., Ferguson, T., and Ghassemi, H. Microbiological aspects of Sangar break, J. Food Sci. Tochnol. (India) 14:251–254 (1977).
19. Barber, S., Tormer, M. J., Martinez-Anaya, M. A., and Benedito de Barber, C. Microflora of the sour dough of wheat flour. IX: Biochemical characteristics and baking performance of wheat doughs elaborated with mixtures of pure microorganisms. Z. Lebensm. Unters. Forsch. 189:6–11 (1989).
20. Vogel, R. F., Boeker, G., Stolz, P., Ehrman, M., Fanta, D., Ludwig, W., Pot, B., Kersters, K., Scheifer, K. H., and Hammes, W. P. Identification of lactobacilli from sour dough and description of Lactobacillus pontis, sp. vov. Int. J. Syst. Bacteriol. 44:223–229 (1994).
21. Dam, W. H. The biochemistry of bakers' yeast—old or new business? In: Chemistry and Physics Baking (T. M. W. Blanshard, P. J. Frazier, and T. Galliard, eds.). Royal Soc. of Chemistry, London, 1986, pp. 117–131.
22. Stauffer, C. E. Leavening agents. In: Functional Additives for Baking Foods (C.E. Stauffer, ed.). Van Nostrand Reinhold, New York, 1990, pp. 189–286.
23. Shalton, D. R. Cereal carbohydrates. In: Handbook of Cereal Chemistry and Technology (K. Kulp and J.G. Ponte, eds.). Marcel Dekker, New York, 2000, pp. 385–386.
24. Brandt, M. J. and Hammes, W. P. Effect of fructosans an soudough fermentation (German). Getreide Mehl und Brot 55:341–345 (2000).
25. Sanderson, G. W., Reed, G., Bruinsma, B., and Cooper, E. J. Yeast fermentation in bread making. Res. Dept. Am. Inst. Baking, Manhattan KS, Tech. Bull. 5(12) (1983).
26. Sanderson, G. W. Yeast products for baking industry. Cereal Foods World 30:710–775 (1985).

27. Jonas, V. Bakers Yeast Technology. Association of Czech Yeast Manufacturers, 1943, pp. 5–12.
28. Thorn, J. A., and Ross, J. W. Determination of yeast growth in doughs. Cereal Chem. 37:415–421 (1960).
29. Triveti, N., Hauser, J., Nagodawithana, W. T., and Reed, G. Update on Baker's yeast. Am. Inst. Baking, Manhattan KS, Res. Dept. Tech. Bull. 11(2) (1989).
30. Carlin, G. T. The fundamental chemistry of bread making. Proc. Am. Soc. Bakery Eng., pp. 56–63 (1958).
31. Fowell, M. S. Infection control in yeast factories and breweries. Proc. Biochem 2(12):11–15 (1967).
32. Doerry, W. Sourdough and breads. Am. Inst. Baking Res. Bull. 20(7) (1998).
33. Ziemke, W. H., and Sanders, S. Sourdough bread. Am. Inst. Baking Res. Bull. 10(10) (1988).
34. Doerry, W., and Hutz, E. Sourdough starters and breads, Am. Inst. Baking Res. Bull. 22(3) (2000).
35. Lugon, and Brescia R. Panettone, Am. Inst. Baking Res. Bull. 14(8) (1992).
36. Kulp, K. Bread industry and processes. In: Wheat Chemistry and Technology, 3rd ed. (Y. Pomeranz, ed). American Association of Cereal Chemists, St, Paul, MN, 1988, Vol II, 371–406.
37. Bogy, T., Affordable artisan bread products. Proc. Am. Society Baking, 169–177 (2000).
38. Sullivan J. M., and Malovany, D. The rustic bread revolution. Bakery Production and Marketing, April 24:18–24 (1995).
39. Forestier, D. Technology of producing French breads (in France and North America). Am. Inst. Baking Res. Bull. 18(7) (1996).
40. Wehrle, K., Grau, H., and Arendt, E. K. Effects of lactic acid, acetic acid, and table salt on fundamental rheological properties of wheat dough. Cereal Chem. 74(6):739–744 (1997).
41. Zobel H. F., and Kulp, K. The staling mechanism In: Baked Foods Freshness (R. E. Habeda and H. F. Zobel, eds.). Marcel Dekker, New York, 1996, pp. 1–64.
42. Schoch, T. Starches in foods. In: Symposium on Foods, Carbohydrates and Their Roles (H.W. Schultz, R. F. Cain, and R.W., Wrolstad, eds.). AVI, Westport, CT, 1969, pp. 394–450.
43. Calvel, R. The Taste of Bread. Aspen Publishers, Gaithersburg, MD, 2001.
44. Koncel, J. A. Rudi's new bakery meets demands of organic bread market. Bakery Management, January 2002, 24–30.
45. Seiffer, G. Stress-free dough technology. Am. Inst. Baking Res. Bull. 24(3), (2002).
46. Malovany, D. Baking baquettes the long way. Bakery Production and Marketing, April 24, 1995, pp. 26–28.
47. Maga, J. A. Bread flavor. CRC Crit. Rev. Food Technol. 5:55–142 (1974).
48. Rothe, M. Aroma of bread. In: Handbook of Aroma Research (German), Akademie Verlag, 1874, pp. 664–922.

49. Schneeweiss R., and Klose, O., eds. Technology of Industrial Bakery Production (German), Ver. Fachbuchverlag, Leipzig, 1981.
50. Megoffin, C. D., and Hoseney, A. C. A review of fermentation. Baker's Dig. 48(6):22–23, 26–27 (1974).
51. Wiseblatt, L., and Zoumut, H. F. Isolation, origin and synthesis of a bread flavor constituent. Cereal Chem. 40:162–169 (1963).
52. Hunter, I. R., Walden, M. K., McFadden, W. H., and Pence, J. W. Production of bread-like aromas from proline and glycerol. Cereal Sci. Today 11(11):493–501 (1966).
53. Frasse, P., Lambert, S., Richard-Molard, D., and Chiron, H. The influence of fermentation of volatile compounds in French bread dough. Lebensm. Wiss. Technol. 26:126–132 (1993).
54. Czerny M., and Schieberle, P. The aroma of wheat sour doughs—effect of four types (German). Getreide Mehl Brot 56:18–22 (2002).
55. Hansen, A., and Hansen, B. Influence of wheat four type on the production of flavor compounds in wheat sourdoughs. J. Cereal Sci. 19:185–190 (1994).
56. Ehrlich, F. The method of degradation of racemic amino acids by means of yeast. Biochem Z. 1:8 (1906).
57. Ayarapaa, T. Formation of phenyl ethyl alcohol from C^{14} labeled phenylalanine, J. Inst. Brew. 71:341–347 (1965).

6

Commercial Starters in the United States

T. Frank Sugihara
Consultant, Tustin, California, U.S.A.

I. INTRODUCTION

With the recent interest and increase in sourdough breads around the country, numerous popular articles have appeared in magazines and books. The various authors seem to feel that the "sourdough starter" originated in southern Europe, more precisely the Basque country. In fact, one of the oldest bakeries in San Francisco was founded more than 100 years ago (during the Gold Rush days) by Basque immigrants. San Francisco sourdough bread has become world famous and is most likely responsible for its spread across the western United States. During the last 10 years, sourdough bread has become popular in many midwestern cities as well as in the northeastern part of the country.

In the San Francisco Bay area, sourdough bread consumption has constituted more than 20% of all bread produced in the area. A major portion of the sourdough bread has been produced by using the natural "mother sponge" or sourdough starter [1–3]. In most cases the procedure is as follows: The sourdough sponge (mother sponge) is generally made up as shown in Table 1. Approximately 100 parts of the previous sponge is mixed with 100 parts of high-gluten flour and 46–52 parts of water. Essentially it contains only flour and water with microorganisms transferred from the previous starter sponge. Of great importance here is the "acidity" or pH

Table 1 Starter Sponge Formulation

100 parts of the previous sponge (40% of final mix)
100 parts of flour (high-gluten flour)
46–52 parts water
Starting pH 4.4–4.5
Final pH 3.8–3.9

Source: Ref. 1.

range of the system. The starter begins with a pH of 4.4 and levels off at a pH of 3.9, an extremely acidic system. The acidity of the system prevents contamination by other microorganisms. It takes about 8 h for the starter to fully develop. If the starter sponge is not used for the day's production, it should be stored at 42–55 °F.

The bread dough (Table 2) is formulated with about 20 parts of starter sponge (11% of final mix), 100 parts of flour (regular patent), 60 parts of water, and 2 parts of salt. No sugars, shortening, non-fat milk solids (NFMS), yeast foods, oxidizing agents, dough softeners, or conditioners are used. After about 1 h floor time, the bread dough is molded and proofed for about 8 h. The pH of the bread dough drops from about 5.3 to 3.9. The bread dough is baked for a relatively long time (45 to 50 min) at a relatively low temperature (375–390 °F) with low-pressure steam introduced during the first half of the baking cycle or until the crust begins to develop color. The use of live steam or a wet oven is of critical importance to the eating quality of the crust, as is the making of razor-like cuts on the upper surface of the dough just before placing it in the oven. Baking is generally done directly on a carborundum hearth. There is little or no pH change during baking, so that the bread has a pH of 3.9–4.0.

Table 2 Bread Dough Formulation

20 parts starter sponge (11% of final mix)
100 parts flour (regular patent)
60 parts water
2 parts salt
Starting pH 5.2–5.3
Final pH 3.9–4.0

Source: Ref. 1.

San Francisco sourdough bread, seems, superficially, easy to make with few ingredients but we must remember that the leavening as well as the souring must be accomplished by microorganisms. The "heart" of the process is a dynamic, living, biological system that requires specific environmental conditions. If the conditions are not right, the system may be invaded by other unfavorable microorganisms or it may just expire and fail to make a loaf of bread. But with proper care, the mother sponge or starter can be perpetuated indefinitely. Some bakeries have perpetuated their starters for more than 100 years.

Researchers at the USDA laboratory [4–6] examined mother sponges (starters) from five different commercial sources for microbial flora. All were found to contain the same rather unusual species of yeast, which was readily classified, in part by its inability to ferment or assimilate maltose and its ability to grow out well in the presence of actidione (cycloheximide), an antibiotic inhibiting the growth of most yeasts (Table 3).

Microscopically, the pure isolates were all typical yeast cells, globulaar to ovoid in shape, and showing "budding-fission" reproduction (Fig. 1). Leavening action was found to correlate well with the presence and numbers of this strain of yeast, occurring at levels of 10–25 million cells per gram of mother sponge or starter. Initial yeast levels in the bread doughs were about 2–4 million cells per gram or roughly 1/50th the level at which baker's yeast is used in conventional bread dough. The maltose-negative aspect of this yeast was very surprising because maltose is the principal fermentable sugar available in doughs prepared with addition of sugars. After rigorous taxonomic procedures, the maltose-negative isolates were identified as asporogenous strains of *Saccharomyces exiguus* (*Torulopsis holmii*), [4].

Studies with pure cultures of *S. exiguus* (sourdough yeast) confirmed its role as the leavening agent in sourdough and demonstrated its unusual vigor and growth in a simulated sour dough environment (approximately 50% of the acidity attributable to acetic acid). Under the same conditions baker's yeast was found to die off (Fig. 2).

Table 3 Fermentation Characteristics of Sourdough Yeast (*Saccharomyces exiguus*)

Yeast	Glucose	Sucrose	Raffinose	Galactose	Maltose	± on Actidone
Baker's	+	+	+	+	+	−
Sourdough	+	+	+	+	−	+

Source: Ref. 1.

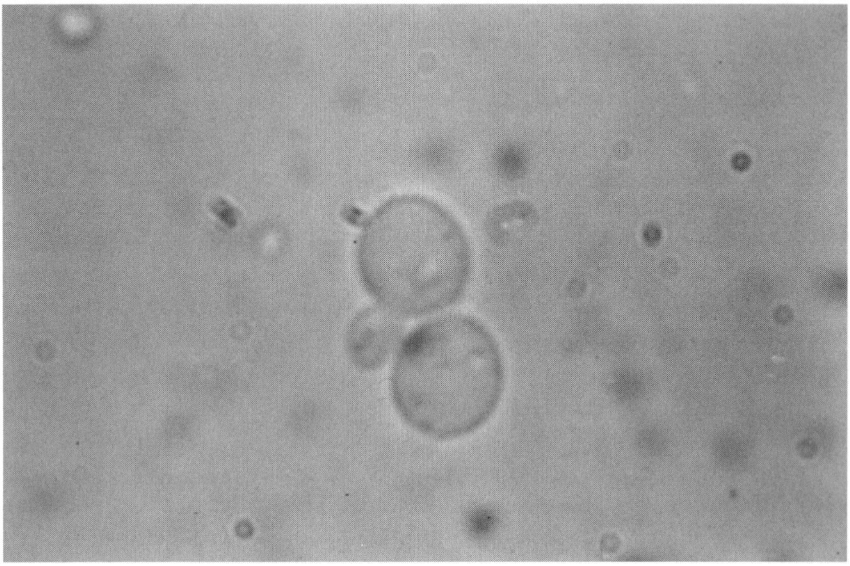

Figure 1 Sourdough yeasts (*Saccharomyces exiguus*) ×1000. (From Ref. 1.)

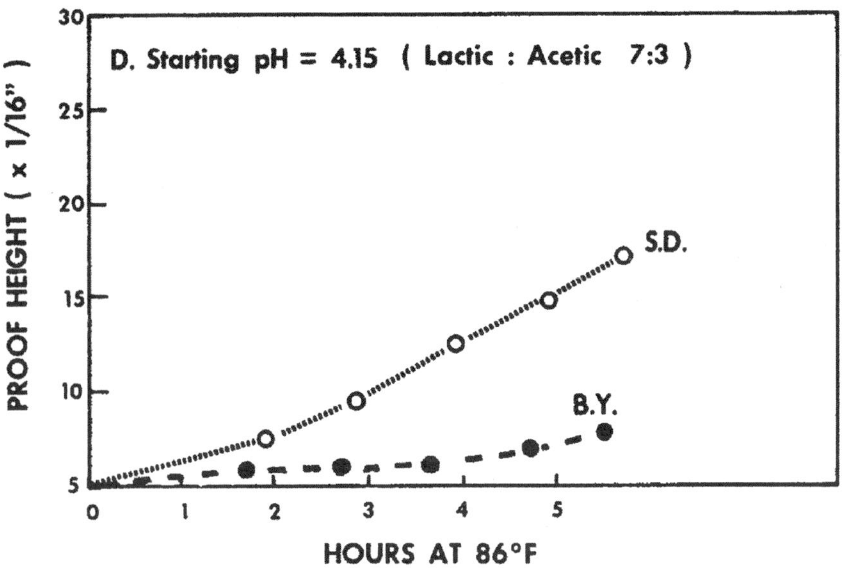

Figure 2 Proofing ability of sourdough yeast (*S. exiguus*) in a simulated sourdough compared to commercial baker's yeast. (From Ref. 1.)

Ng [7] studied the nutritional requirements of the San Francisco sourdough yeast, *S. exiguus*. After extensive studies, he found that *S. exiguus* can be grown in substantial quantities in a simple molasses (10 Brix) medium with a added source of thiamine. Unfortunately, *S. exiguus* is not stable in compressed cake form as in the case of baker's yeast. Other forms that were studied, such as flash freezing in a stabilizer and freeze drying, were not feasible.

A direct microscopic examination of the starter using staining techniques showed us beyond doubt that a bacteria was also involved in the process (Fig. 3). Initial attempts to isolate bacteria of any type in appreciable numbers from sourdough mother sponge or starter, using several dozen plate count, lactic, acetic, anaerobic, coliforms, etc. were unsuccessful. The use of agar media alone or agar supplemented with flour and yeast extractives was also unsuccessful. Eventually, it was found that an unusual combination of nutritional requirements, including an absolute requirement for maltose, was necessary to obtain growth on isolation media. Fortunately, this requirement for maltose dovetails, with the finding that the sourdough yeast (*S. exiguus*) doesn't utilize maltose. Heavy growth of these bacteria in 1–2 days was then achievable and permitted their enumeration. Bacterial counts in the various mother sponges ranged from 1/2 to 1-1/2

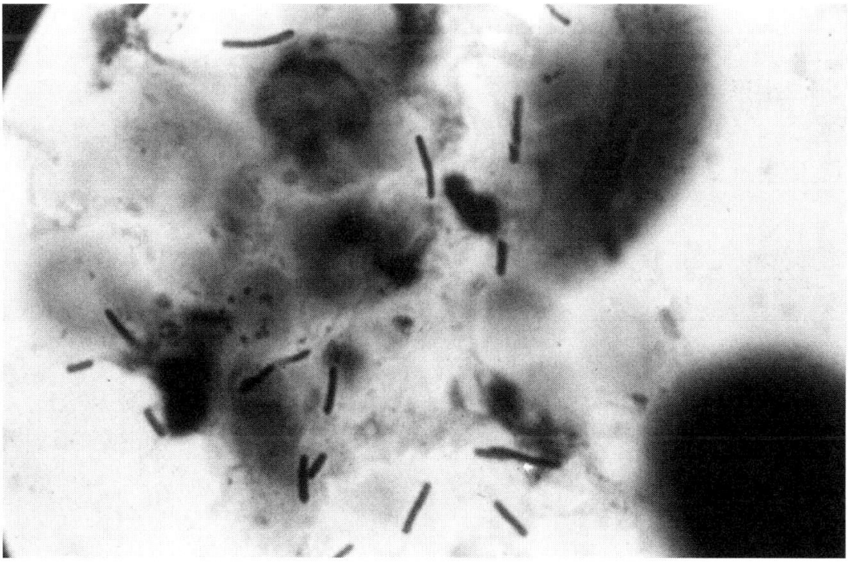

Figure 3 Sourdough starter, showing yeast and lactic bacteria ×1000. (From Ref. 1.)

billion cells per gram, or roughly 50–100 times more numerous than the sourdough yeast cells. The bacteria were similar or identical from the various sources and morphologically varied in appearance from short rods to involuted, swollen, and filamentous forms all derivable from a pure culture (Figs. 4 to 6). These bacteria did not appear to fit any known taxonomic position. Kline and Sugihara suggested the name *Lactobacillus sanfrancisco* in honor of their unique role in this uniquely situated process [5].

Additional taxonomic studies were conducted by Sriranganathan et al. [8]. They found that the guanine plus cytosine base composition (% GC) of the desoxyribonucleic acid (DNA) ranged from 38% to 40%. The possible genetic relatedness of these sourdough isolates to five known species of *Lactobacillus* with comparable GC contents was assessed by means of DNA-DNA hybridization competition experiments. The sourdough bacteria exhibited a high degree of homology (more than 88%) among themselves, suggesting that the four isolates studied were identical taxinomically. Also, the electron photomicrographs revealed structures similar to those of gram-positive bacilli. The sourdough isolates were found to have the characteristic phenotypic and morphological properties of the genus *Lactobacillus* and are not related genetically to any known species, the tentative characterization by Kline and Sugihara of these isolates as a new species is substantiated.

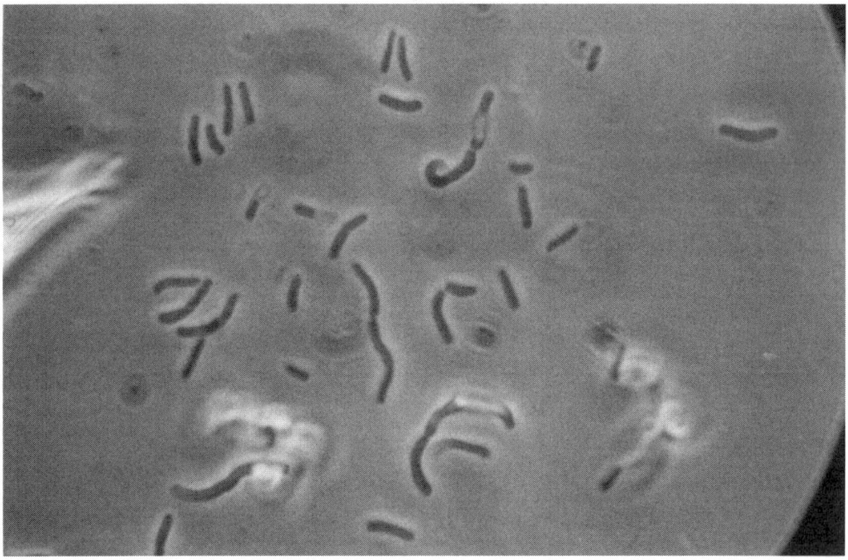

Figure 4 Sourdough bacteria (*Lactobacillus sanfrancisco*) $\times 1000$. (From Ref. 1.)

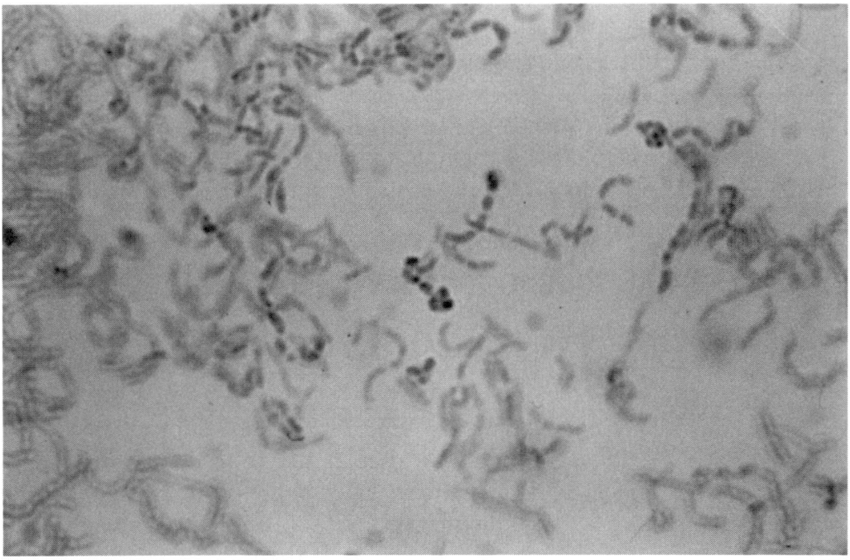

Figure 5 Sourdough bacteria (*L. sanfrancisco*) showing involuted and swollen forms, ×1000. (From Ref. 1.)

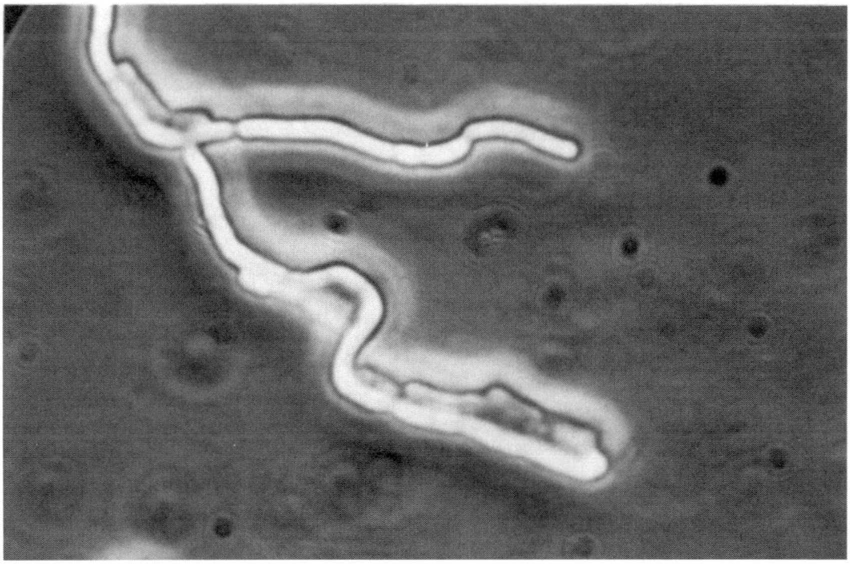

Figure 6 Sourdough bacteria (*L. sanfrancisco*) filamentous forms, × 2000. (From Ref. 1.)

Starter cultures of the San Francisco sourdough bacteria, *L. sanfrancisco*, can be produced in two forms. It can be grown in a pure culture using a chemically defined media or grown in a essentially pure form on a wheat flour and water substrate [9]. Both techniques are being used today for commercial bakery starters. Both starters must use baker's yeast in the final dough because sourdough yeasts are not available commercially.

II. COMMERCIAL SOURDOUGH ADDITIVES

There are two types of additives available for the production of sour dough bread:

1. *Chemical*: lactic acid, acetic acid, or a combination of both and usually sold encapsulated for ease in adding to dough mixture.
2. *Dried fermented flour*: a flour water slurry is fermented with a lactic bacteria and then dried. This product is added to the formula usually at a rate of 5–15% of the total flour.

These additives are not starters (biologically active). They are flavor enhancers.

III. COMMERCIAL SOURDOUGH STARTERS

There are two types of sour dough starters used in the United States: (1) pure cultures of lactic bacteria such as *Lactobacillus sanfrancisco*, *Lactobacillus brevis*, *Lactobacillus plantarum*, *Lactobacillus leichmannii*, etc.; (2) freeze-dried flour containing biologically active lactic bacteria (Appendix).

Pure cultures are generally grown in chemically defined media such as the sourdough bacteria (SDB) broth or agar as described by L. Kline and T. F. Sugihara [5]. It contained the following: maltose, 2.0%; yeast extract (Difco), 0.3%; fresh yeast extractives (FYEs), 0.5–1.5%; Tween 80, 0.03% (3 mL of 10% solution/L); and trypticase (BBL) 0.6%. The pH was adjusted to 5.6 with 20% lactic acid or 1–6 N HCl. The FYEs were usually prepared by autoclaving a 20% suspension of commercial compressed baker's yeast in distilled water for 30 min at 15 psi, allowing the suspension to settle overnight at 2–8 °C, decanting, and further clarifying the supernatant by centrifugation. The extract prepared in this manner contained 1.5% solids and, if not to be used within a few days, was frozen immediately.

For the production of commercial quantities of *L. sanfrancisco*, the SDB broth is inoculated from an SDB agar slant in increasing volumes in fermentors. After growing out a suitable quantity, the SDB both laden with cells of *L. sanfrancisco* is centrifuged to separate the cells from the broth. The cells are then suspended in a stabilizing medium consisting primarily of skim-milk solids. This mixture is then freeze dried and packaged in sealed envelopes. These pure cultures are packed in quantities suitable to inoculate a culture dough or sponge, which is fermented for 16–24 h before mixing into the final dough. Baker's yeast must be added to the final dough since the lactic culture does not contain the sourdough yeast found in mother sponges. Final proof time depends on the quantity of baker's yeast added. To produce a typical San Francisco sourdough bread, which generally has a proof time of 7 h, 0.2% baker's yeast is added to the final dough. Additional information on the production of *L. sanfrancisco* starter culture can be found in two U.S. Patents by Sugihara and Kline [10,11].

IV. STARTERS FOR THE PRODUCTION OF SODA CRACKERS

Soda crackers have been produced for decades by a 24 h process. The process relied on the dough's adhering to the sides of the mixing trough for the inoculum, which was seldom clean. To accelerate the fermentation, a small portion of the previous dough is usually added. A few microbiologial and chemical studies have been published by Mika [12–14] but the microorganisms responsible for the process have never been isolated or identified.

Studies were conducted by Sugihara [15,16] on the isolation of microorganisms from soda cracker doughs supplied by two commercial plants over a period of 6 months. The samples were analyzed as follows: an 11-g sample was aseptically blended for 90 s at a reduced speed with 99 mL of sterile 0.1% aqueous peptone in a sterilized 8-oz. Osterizer jar. After additional serial dilution, also in sterile 0.1% peptone, samples were plated on selected agar media by the spread-plate technique and incubated for 2 days at 30 °C. Preliminary media screening indicated that the only suitable media were tomato juice agar (Difco Laboratories, Detroit, MI) and MRS agar (Difco). They were selected for visual ease in colony counting and reproducibility of the media. Samples from the two commercial plants were received once a month for 6 months to study the possibility of changing microflora. Samples were shipped in insulated boxes with frozen refrigerant (canned gel), air freighted, and delivered within 24 h.

Several hundred isolates were collected for screening. Single colonies from spread plates were picked and carried on MRS slants. The yeasts were eliminated from the screening since the standard formula includes 5×10^6

baker's yeast (*S. cerevisiae*) per gram of dough. Contaminating yeasts would be few.

Biochemical studies were made using a modified MRS medium. Tween 80, 1.0 g; ammonium citrate, 2.0 g; sodium acetate, 5.0 g; manganese sulfate, 0.05 g; and disodium phosphate, 2.0 g/L distilled water. The pH was adjusted to 6.6 with HCl. Bromthymol blue (0.1 g/L) was added to give a dense green color. Carbohydrate solutions (6%) were made and sterilized by passing through a 0.02-μm filter membrane. Fermentation studies were conducted in 20 × 125 mm screw-capped culture tubes with a 10 × 44 mm inverted culture tube insert for gas collection. Culture tubes, with inserts, and 8 mL of basal media were sterilized at 15 psi steam for 15 min. After cooling, 4 mL of sterile carbohydrate solution was added.

The inoculum was prepared by suspending the growth of 24-h-old MRS slant in 5 mL of sterile distilled water. Tubes were inoculated with 0.1 mL of the suspension and incubated at 30 °C. Metabolic data were recorded at 24 and 48 h. *Bergey's Manual of Determinative Bacteriology* was used for determining species of lactic bacteria isolates. Approximately 200 bacterial isolates were examined in this study. All were found to be gram-positive, catalase negative, and nonmotile rods. After analysis of the carbohydrate reactions of the isolates, they were compared with data in *Bergey's Manual* which indicated that the dominant species was *L. plantarum*. Two other species were also found in substantial numbers, i.e., *L. delbrueckii* and *L. leichmannii*. Smaller numbers of *L. brevis*, *L. casei*, and *L. fermentum*. Species distribution found in the soda cracker dough are shown in Table 4.

Pure cultures of the three most dominant species were grown and tested on the fermentation of the soda cracker process. Culture concentrates

Table 4 Species Distribution of Lactic Bacteria Isolated from Soda Cracker Sponge and Dough

Species	% of Isolates
Lactobacillus plantarum	47
Lactobacillus delbrueckii	16
Lactobacillus leichmannii	13
Lactobacillus brevis	5
Lactobacillus casei	4
Lactobacillus fermentum	3
Other lactic bacteria	12

Source: Ref. 13.

Table 5 Survival of Viable Cells in Frozen Concentrates After Storage (10 months at −100 °C)

	Counts/g	
Species	"0" Time	10 Months
L. plantarum	10×10^{10}	8×10^{10}
L. leichmannii	17×10^{10}	8×10^{10}
L. delbrueckii	15×10^{10}	8×10^{10}

Source: Ref. 13.

with viable counts of 10–17 $\times 10^{10}$ were prepared, frozen, and stored at −100 °C. All showed excellent frozen stability properties. Viable counts remained constant for at least 10 months (Table 5).

For ease of handling and economics, a "liquid sponge" step was developed. The liquid sponge when used to inoculate the conventional sponge gave a starting lactic bacteria count of 1×10^8. The conventional lag phase was eliminated. Sponge fermentation was completed in 4 h (conventional sponge fermentation takes 18–20 h). Dough fermentation was also rapid, i.e., 2 h instead of the usual 4 h (Table 6). It is conceiveable to schedule an 8-h production run using pure culture technology.

Soda crackers were made on a laboratory scale to determine the quality of the product using pure culture fermentation. *L. plantarum*, *L. delbrueckii*, and *L. leichmannii* were used singularly. For mixed-culture fermentation studies, *L. plantarum*, the most prevalent species, was used with either *L. delbrueckii* or *L. leichmannii*. Fermentation time and physical characteristics of the final product were essentially identical. There were

Table 6 Shortening the Fermentation Time for the Production of Soda Crackers by the Use of Pure Cultures

	Fermentation Time (h)	
Stage	Conventional	Pure culture
Liquid sponge	0	4
Sponge	18	4
Dough	4	2
Total time	22	10

Source: Ref. 14.

slight differences in flavor when tested organoleptically. Mixed cultures were preferred by most judges in the test. Differences in flavor is probably due to organic acid composition and quantities of diacetyl present.

Pure culture fermentation in the production of soda cracker was tested on a commercial scale and shown to be feasible. Pure cultures are available in commercial quantities.

V. COMMENT

The author has devoted most of this chapter on the lactic starter cultures commonly used in the United States. There are many important lactic starter cultures manufactured and used in Europe, such as *Lactobacillus brevis* ssp. *lindneri*, for rye bread production and other lactic cultures used for production of Italian panettone, pandoro, and other ethnic bakery products. These cultures are discussed in other chapters of this book.

APPENDIX

Manufacturers and Distributors of Starter Cultures

This is a tentative list of suppliers offering biologically active bakery starters. Some are manufactured in the United States and others made in Europe and distributed here.

Name	Bacteria
Lyoferm Inc, 3862 E. Washington St. Indianapolis, IN 46201 (317) 359–9528	*Lactobacillus sanfrancisco* *Lactobacillus plantarum* *Lactobacillus brevis* *Lactobacillus delbrueckii*
Chr. Hansen's Lab., Inc. 9015 W. Maple St. Milwaukee, WI 53214 (414) 476–3630	FloraPan L-62 (*Lactobacillus brevis*) FloraPan L-73 (*Lactobacillus plantarum*)
Lallemand, Inc. 1620 Prefontaine Montreal, Quebec H1W2N8 (514) 255–7747 Ext. 231	Bakery Starter Culture Type I *Pediococcus acidilacti* & *Saccharomyces cerevisiae* Bakery Starter Culture Type II

Name	Bacteria
	Pediococcus pendtosaceus & *Torulopsis delbreucki*
	Bakery Starter Culture Type III
	Lactobacillus plantarum & *S. cerevisiae*

REFERENCES

1. Sugihara, T. F. Report of Sixth National Conference on Wheat Utilization Research, Agricultural Research Service, U.S.D.A., ARS 74–54 p. 85 (1970).
2. Kline, L., Sugihara, T. F. and McCready, L. B. *Baker's Digest* 44(2):48 (1970).
3. Sugihara, T. F., Kline, L. and McCready, L. B. *Baker's Digest* 44(2):51 (1970).
4. Sugihara, T. F., Kline, L. and Miller, M. W. *Appl. Microbiol.* 21(3):456 (1971).
5. Kline, L. and Sugihara, T. F. *Appl. Microbiol.* 21(3):459 (1971).
6. Sugihara, T. F. and Kline, L. *J. Milk Food Technol.* 38(11):667 (1975).
7. Ng, H. *Appl. Environ. Microbiol.* 31(3):395 (1976).
8. Sriranganathan, N., Seidler, R. J., Sandine, W. E. and Elliker, P. R. *Appl. Microbiol.* 25(3):461 (1973).
9. Kline, L. U.S. Patent No. 4,243,687 (1981).
10. Kline, L. and Sugihara, T. F. U.S. Patent No. 3,734,743 (1973).
11. Sugihara, T. F. U.S. Patent No. 3,891,773 (1975).
12. Mika, J. *Cereal Chem.* 18:483 (1941).
13. Mika, J. *Food Ind.* 20(1):102 (1948).
14. Mika, J. *Cereal Chem.* 32:125 (1955).
15. Sugihara, T. F. *J. Food Prot.* 41(12):977 (1978).
16. Sugihara, T. F. *J. Food Prot.* 41(12):980 (1978).

7

Rye Bread: Fermentation Processes and Products in the United States

Klaus Lorenz
Colorado State University, Fort Collins, Colorado, U.S.A.

I. INTRODUCTION

Rye was brought to North America and western South America with the settling of these areas by Europeans in the 16th and 17th centuries (Bushuk, 1976). The pilgrims brought rye, barley, and oats with them, but no wheat. They ate rye breads. The consumption of rye and rye breads, however, decreased over the years as wheat and white bread became available. Decreases in rye bread consumption were also due to inherent difficulties in producing acceptable rye breads. Good-quality wheat bread can be produced in large mechanized bakeries, whereas good rye bread still must be baked by an expensive, small-scale process (Pomeranz and Shellenberger, 1971). White bread attained great prominence in the 1930s through the 1960s, when it became increasingly fashionable to malign white bread as unappetizing and of low nutritional value. Thereafter, many factors led to the increasing consumption and interest in variety breads, including rye breads. Certainly, one factor was the negativism about American white bread, which was often carried to the extreme by the nutritionally uninformed. It was also aided by the "natural" food movement. We also saw changes in preferences by consumers for the distinctive eating qualities and flavors that characterize different bread types, the increasing perception of the nutritional value of different bread types, changes in the ethnic makeup of broad segments of the population, and, finally, the very natural human appreciation of variety for its own sake (Loeb, 1981).

The continuing and growing popularity of rye breads today rests largely on their distinctive flavor, taste, and eating quality. Rye products offer the baker an opportunity to exercise his or her ingenuity for developing varieties with distinct crust and crumb characteristics, textures, and product forms (Gordon, 1970). A good loaf of rye bread can be a lead item for increasing sales of other bakery products. It gives a definite advantage to the retail bake shop (Sultan, 1965).

The rye bread produced in the United States today is quite different from that baked by the pilgrims. The whole-rye bread has been largely replaced by mixed wheat–rye breads despite the insistence of connoisseurs that the replacement is inferior in both taste and shelf life. Much research has been done to determine optimal conditions of the main phases of bread production such as mixing, fermentation, and baking (Drews and Seibel, 1976). However, the oldest, and still the most important, fermentation process in rye bread production—the sour dough process as it is used by European bakers—is employed for only a few of the rye breads available in the United States. This and the relatively large percentage of wheat flour used in rye breads are the main differences in production of breads in the United States and Europe (Lorenz, 1987).

II. CONSIDERATIONS IN RYE BREAD PRODUCTION

Until recently, many U.S. bread bakers concentrated on white pan bread production because rye breads represented only a small segment of the market. With greater popularity of rye breads, bakers had to change production and consider a number of factors (Ponte, 1981):

1. *Production rates.* Many United States bakers produce white bread in relatively long runs, followed by different variety breads in shorter runs. Frequent changeover from one variety of bread to another causes loss of production time and increases costs. Careful scheduling is required considering personnel availability, mixing requirements, equipment availability, proof times, oven temperatures, baking times, and type of wrapping required. Rye bread production requires some hand labor.
2. *Ingredient costs.* Ingredient costs are sometimes higher for rye breads than they are for white bread. A greater number of ingredients involves a larger inventory with accompanying storage problems.

3. *Processing equipment.* Since rye bread doughs are stiff, processing equipment must be sturdy enough and have suitable longevity. More plant space may be needed to house the equipment.
4. *Packaging.* Rye bread might be wrapped in different and more expensive packaging material. The breads may also be more difficult to slice because they are denser than white bread.
5. *Shelf life.* The shelf life of some rye bread varieties may be limited causing a distribution problem for the baker. These breads are often promoted as "natural, containing no preservatives" and do not contain antimycotic agents.

These factors indicate that the cost of producing rye breads are greater than the costs of producing conventional white bread (Ponte, 1981).

III. DOUGH PROCESSING CHANGES

Rye doughs are processed by methods similar to those used to make white bread but some differences exist (Ponte, 1981).

1. *Formulation.* Many kinds of different ingredients may be added to rye bread doughs. Flour requirements are more demanding because the wheat flour must support considerable quantities of non-gluten-containing materials.
2. *Absorption.* Rye doughs are usually much stiffer than white bread dough. Such stiffness is needed to restrict dough flow and maintain desired product symmetry. Also, many rye breads are marketed as "natural" food products. Therefore, the baker must avoid using additives that keep dough surfaces dry. This requires a lower absorption to avoid dough stickiness and permits optimal handling with dough processing machinery.
3. *Mixing.* Rye dough made with appreciable amounts of nongluten materials requires reduced and/or low-speed mixing. Doughs are readily overmixed producing deleterious effects on product quality.
4. *Fermentation.* Doughs with appreciable amounts of nonwheat flour components are often given a shorter fermentation than white bread doughs.
5. *Proofing.* Rye doughs usually receive a shorter proof time than white bread doughs.
6. *Baking.* The denser rye breads are baked longer at a lower temperature than white breads to ensure good loaf characteristics and maximal flavor development.

7. *Cooling*. Denser rye breads require a longer cooling period than white bread to reach acceptable wrapping temperatures. Slicing of these loaves may be more difficult than slicing of white bread.

IV. RYE BREAD TYPES

In the United States, rye breads are produced in a large number of varieties to meet the specific demands of local consumer preferences. Rye breads vary in crumb color from practically white to a very dark color, in shape from round to elongated, and in taste from a mildly sour flavor to a strong, distinctive acid taste spiced with caraway seed. Rye breads may contain only the basic ingredients—wheat and rye flours, water, yeast, and salt—or may incorporate a number of additional ingredients, such as molasses, potato flour, sugar, shortening, buttermilk, nonfat dry milk, all designed to enhance either the flavor, color, or keeping quality of the bread. It is therefore possible for the baker to modify the rye bread formula to produce the type of product that will find greatest appeal among his customers.

The many types of rye breads undoubtedly have evolved to meet local preferences, and many descriptive names (American, Jewish, German, Swedish, Black Forest, Bohemian) are used. Those breads are nonstandardized, so that in the United States they may vary considerably depending on where they are produced (Ponte, 1981; Lorenz, 1980, 1982, 1987).

As is true of all quality products, the production of rye bread of optimal flavor and eating properties requires that the baker exercise good judgment and proper care in selecting ingredients and in the processing to which these ingredients must ultimately be subjected. The basis character of rye bread is determined by the type of rye flour used and the amount or proportion incorporated in the dough formula. It is important, therefore, to distinguish between the various types of rye flour available to the baker and to recognize the differences that exist between them (Meyer, 1965; Lorenz, 1982).

V. RYE BREAD INGREDIENTS

A. Rye Flour

It is generally known that rye flour does not contain the type of protein system, characteristic of wheat flour, that leads to gluten formation during mixing. However, this does not mean, that the differences between various types of rye flour are without significance.

Rye flours may be classified into four general types: (1) white or patent grade; (2) medium or straight grade; (3) dark grade; and (4) meal or whole grain.

White or patent rye flour is a product that is essentially white and relatively bland. It is composed almost entirely of the endosperm portion of the rye grain, i.e., mainly starch, which exerts a minimal loaf volume depressing effect. It finds extensive use in the production of Jewish-type rye bread. When used in rye bread production, it can exceed 40% in the total flour blend without noticeably reducing the loaf volume.

Medium or straight grade rye flour comprises all flour that can be milled from the rye kernel, excluding, of course, the bran and germ portions, and corresponds in grade to the straight flour of wheat milling. It possesses a somewhat distinctive and characteristic rye flavor. It can be used in relatively large proportions in rye bread production and gives good results. Up to 30% of the total lbs of flour may be used to produce an excellent and flavorful loaf of rye bread with a highly acceptable volume.

Dark grade rye flour represents the coarser portion of a straight rye flour that remains after the white or patent flour has been removed. It is distinctly a lower grade flour, relatively dark and with a pronounced rye flavor. It is well suited for the production of Pumpernickel-type rye breads, imparting to them their dense grain and full flavor, but when used in rye bread its proportion must be held to below 15% in the total flour blend; otherwise loaf volume will suffer excessively. In general, the dark flours possess higher absorption and moisture retention capacities than do the light grades of rye flour.

Rye meal, also known as Pumpernickel flour, is obtained by grinding of whole rye kernel. It is available in a range of granulations, depending on the degree of grinding. The more finely ground grades are readily adapted to rye bread production, with the proportions used corresponding essentially to those of the dark rye flour.

By judiciously blending various proportions of these different rye flours, the baker is able to produce rye breads of distinctive and unique taste appeal (Meyer, 1965; Lorenz, 1982, 1987). Product specifications of rye flours are given in Table 1.

Some mills will produce other grades intermediate between the three grades given above. An extra dark rye is sometimes made that will have an ash content as high as 3.0%. Most of the major mills market rye blends that are made from a mixture of strong spring wheat clears and rye. These will roughly follow a pattern of 70% clear and 30% medium rye, 80% clear and 20% dark rye, and 50% clear and 50% white rye. Other blends may also be custom blended to suit a customer's own market need and the rye bread he or she desires. The characteristic rye flavor increases as the ash increases and

Table 1 Basic Grades and Product Specification of Rye Flours

	Moisture (%)	Ash (%)	Protein (%)	Color
White or light rye flour	14.5 maximum	0.58–0.78	7.0–9.1	White
Dark rye flour	14.5 maximum	2.05–2.83	13.7–16.2	Dark
Medium rye flour (blends of above 2 grades)	14.5 maximum	1.11–1.39	10.1–12.8	Medium white
Rye meals (may be flaky, coarse, medium, fine, extrafine)	n.s.	n.s.	n.s.	n.s.

Source: From Lorenz (1982, 1987).
n.s.—no specification.

the color deepens; therefore, if one wishes to impart a greater rye flavor to a product, the darker rye flours should be used (Schmalz, 1970).

Entire rye grain products consist of rye meals (sometimes referred to as rye chops and sometimes as Pumpernickel), flattened rye grain (referred to as flaked rye), and sharply cut rye (called steel cut cracked rye). There is no standardization either in the terminology or the granulation of these above products, and a great deal of confusion exists in the trade in communicating between buyer and seller for the correct designation of the desired product. Usually, the buyer submits a sample to the miller for matching to get the desired product. A great service could be rendered to the rye milling industry if the manufacturers would agree on standards of granulation and nomenclature.

In the area of rye meals, which are analogous to whole wheat, granulations from extrafine to extracoarse and rye flakes from thick to thin are available. The meals are produced on a very short system mill similar to a whole wheat mill (Schmalz, 1970). Uses of the different rye flours in various rye breads are indicated in Table 2.

While differences in color, in dependence on extraction rates, are readily apparent in rye flours, a property of equal importance to ultimate bread quality, but one that is less easily detected and evaluated, is freshness. It is axiomatic of rye flour that for any given grade the best product is the freshest, in contrast to white wheat flour where extended aging is frequently desirable and even necessary for optimum baking performance. Thus,

Table 2 Basic Rye Flours: Types and Uses

Rye flour type	Suggested uses
White rye	Jewish and other rye breads Dusting flour (bench type) Light Swedish rye bread
Medium rye	Bohemian, Polish, and Russian rye breads For blending with other rye flours In American-type rye breads Medium Swedish rye breads Base for rye sours
Dark rye	American rye breads (lower levels than medium rye flour) Dark, heavy German rye breads Pumpernickel breads—for flavor and color Swedish rye bread Base for stronger rye sours
Rye meal (various granulations)	In various types of pumpernickel bread For dusting in hearth rye breads
Cracked rye	Used in either white bread dough or rye to make a specialty cracked rye bread (should be soaked before adding to dough)
Rye flakes (flat, rolled whole rye)	For blending with meal and rye flour in pumpernickel bread In crisp, flat Swedish rye bread In rye crisp wafers

Source: From Gordon (1970) and Lorenz (1987).

regardless of the grades of rye flour a baker elects to use, he must make certain that the flour is fresh if he is to obtain a baked loaf of the desired flavor and keeping quality (Meyer, 1965).

B. Wheat Flour

As far as the wheat flour in rye bread production is concerned, a strong first clear is generally recommended. The blending of wheat and rye flours is necessary to produce a loaf of the desired volume since rye flour lacks the gluten capable of retaining the gases formed during fermentation. Strong patent and second clear flours can be used successfully in a blend. The stronger the wheat flour, the greater the percentage of rye flour that can be used without a reduction in loaf volume.

C. Various Other Ingredients

The typical rye dough contains, in addition to blends of wheat and rye flours, a number of other ingredients in smaller amount (Meyer, 1965; Lorenz, 1987).

1. Salt

Salt is used at slightly higher levels in most rye breads than normally used for white bread. Rye flour is more fermentable than wheat flour. Therefore, salt, plays a more significant role as a fermentation regulator. Salt also greatly enhances the true rye flavor of rye breads.

2. Malt

Malt is an optional ingredient. It stimulates fermentation and imparts a desirable color and a good sheen to the crust. If excessive amounts are used, very sticky doughs may result and the bread could have a dense, gummy crumb.

3. Shortening

Shortening plays an important role in rye bread production, especially if all of the makeup is done mechanically. It also helps to improve keeping quality, crust tenderness, and slicing operation. Excessive gumming and smearing of slicer blades can be minimized by an adequate level of fat in the formulation.

4. Yeast Food

Requirements for yeast food in rye doughs are somewhat lower than in white bread doughs, since one of its principle functions is to stimulate fermentation activity and this is already adequate in rye doughs.

5. Sugar

Sugar is used to impart sweetness and improve crust color. Sucrose, brown sugar, or molasses may be used.

6. Milk

Nonfat dry milk or buttermilk solids may be used to improve texture, crust and crumb color, slicing performance, and nutritional value (Silva, 1956).

7. Yeast

Yeast is the primary leavening agent in mixed wheat–rye breads. The percentage used will depend on the amount of rye flour in the formulation and the desired duration of fermentation.

VI. SOURS AND CULTURES

While most of the rye breads in the United States are manufactured without a sour, there are some rye bread types that require a sour to produce desired flavor and texture characteristics.

Sourdoughs perform three functions:

Leavening
Acidification
Flavor development

Preparation of a sour requires knowledge of the biological principles of the proliferation of bacteria and yeasts. The baker must be willing to take the necessary care to control conditions under which the ferments are being prepared.

All sour-type doughs use a starter in the initial stage. The starter is taken from a ripe sour. It can also be produced by "spontaneous fermentation," a process that is initiated by bacteria indigenous to the flour and by airborne organisms infecting the flour. When a dough undergoes spontaneous fermentation and is maintained for several days, frequent "refreshing" by adding flour and water is required (Spicher and Stephan, 1993; Pomeranz and Shellenberger, 1971). Changes in such environmental dough conditions as temperature, pH, viscosity, and nature of available nutrients can exert profound influences on the character of the microbial flora present. These factors make it possible to control and regulate the quality of the final products.

Depending on whether lactic acid bacteria or acetic acid bacteria gain predominance, the flavor of the sour and of the resultant bread will assume a different and distinctive character. Maximal acid formation occurs around 38 °C. In a soft dough, bacterial growth and acid formation are enhanced; in a stiff dough, acidification is reduced. In soft doughs fermented at about 30 °C and containing up to 2% added compressed yeast, acidification is markedly reduced. Acidification is considerably retarded by the addition of salt to rye sours. At high levels of salt (up to 5% on a flour basis), fermentation can be prolonged.

The microbial flora can undergo a gradual or rapid change in the course of a few days or a week, thereby altering the flavor characteristics of the bread. Obviously, product uniformity under such conditions can hardly be maintained (Spicher and Stephan, 1993). To overcome these problems, commercially prepared sours and cultures, either in dry or wet form, have been introduced. Their advantages include greatly improved uniformity, a relatively strong flavor character of the desired quality, and the fact that with their use the proportion of rye flour in the dough formula can be reduced without loss of flavor intensity.

The dry sours are generally used at a level of 2–4% although it is well in all instances to follow the manufacturer's recommendations. Availability and sources of these commercial sours is the subject of a separate chapter in this book.

The most common method of rye sour production in U.S. bakeries is the so-called multiple-stage or progressive method. It involves producing several sours before the final dough is mixed. Main acidification occurs in the basic sour that is held overnight. The necessity of "refreshing" the sour, the large amount of labor, and the difficulty in controlling the individual stages have drastically reduced the manufacture of bread by that method. A typical sourdough production process is given in Table 3. Other sourdough processes, mostly used in Europe, are discussed by Drews and Seibel (1976).

Undeniably, taste and flavor of bread can be improved by the optimal use of sourdough. In the dough and fermentation stages, precursors of aromatic substances are formed. During baking, they turn into aroma components, especially in the crust. Although the baking process causes many aromatic substances to evaporate, considerable amounts will remain in the bread (Rohrlich, 1961; Brümmer, 1991; Seibel and Brümmer, 1991).

VII. DOUGH METHODS

While both the sponge dough and the straight dough methods, each with or without the use of a sour, may be used in rye bread production, the sponge-and-dough method is preferred by the majority of bakers (Pyler, 1973). This method has been shown to yield:

Better grain and texture
Better keeping qualities
Greater uniformity
Greater fermentation tolerance
Slightly larger loaf volume
Better machinability of the dough

Table 3 Production of a Sourdough Starter

Step	Dough from previous step		Dark rye flour or rye meal		Water		Total weight		Temperature		Time of ripening (h)
	Grams	Pounds	Grams	Pounds	Grams	Pounds	Grams	Pounds	°F	°C	
Starter	—	—	500	1	1,000	2	1,500	3	79–82	26–28	24
Anfrischsour	1,500	3	1,500	3	1,500	3	4,500	9	79–82	26–28	24
Basic sour	4,500	9	4,500	9	4,500	9	13,500	27	90–95	32–35	8
Full sour	13,500	27	13,500	27	13,500	27	40,500	81	79–82	26–28	16

Souce: From Lorenz (1983, 1987).

The straight dough method will also produce good results but requires more careful supervision (Pyler, 1973).

If a pronounced rye flavor is desired, best results are obtained when sponges are made of all rye flour. Absorption should be adjusted to yield a relatively soft sponge. The absorptions of the various types of rye flour differ considerably. In an all rye sponge, a dark rye flour will have an absorption of about 200%, a medium rye flour 150%, and a light rye flour 100%. Depending on the type of rye bread manufactured, and also on the type of rye flour used, either all, part, or none of the rye flour may be used in the sponge. The amount of rye flour in the sponge will affect fermentation rate. Dark rye flours, because of their greater acidity and possibly higher amylolytic activity, mature faster than do light rye flours (Doose, 1964; Pyler, 1973).

VIII. RYE BREAD FORMULATIONS

Rye bread formulations for breads made by the sponge-and-dough procedure and manufactured without the use of a sour are given in Table 4. Rye bread formulations for breads made with a sour dough are given in Table 5.

IX. DOUGH PROCESSING

A. Mixing

The mixing process for wheat and rye doughs differs. Rye doughs are generally mixed at low speed for a long time. However, it is also possible to mix rye doughs at high speed with excellent results. Just what type of mixing treatment is correct for a rye dough depends primarily on the kind and amount of rye flour used and on the strength of the wheat flour. In practice, the greater the proportion of rye flour in the formula and the darker the rye flour, the slower should be the mixing speed and the shorter the mixing time. Dough temperatures are 78–82 °F or 25–28 °C for mixed rye breads and 84–87 °F (28–30 °C) for all rye breads (Slaten, 1960; Lorenz, 1987).

B. Fermentation

While the optimal pH for fully fermented wheat doughs is between 5.0 and 5.3, the optimal pH value for rye doughs is between 3.5 and 4.5. The optimal

Table 4 Rye Bread Formulations, %[a] (No Sour Used)

	American type								
Ingredient	Light Rye	Medium Rye	Dark Rye	Swedish Rye	Bohemian Rye	Jewish Rye	Cheese Rye	Raisin Rye	Salty Rye
Sponge									
Flour—clear	70	70	70	66	66	—	70	30	—
—patent	—	—	—	—	—	40	—	30	60
—light rye	—	—	—	—	—	20	—	—	—
—dark rye	—	—	—	—	—	—	—	—	—
Water[b]	60	60	60	63	63	62	63	64	62
Yeast	2.5	2.5	2.5	2.5	2.5	2	2.5	2	3
Mineral Yeast Food	0.5	0.5	0.5	0.5	0.5	0.25	0.25	0.25	0.5
Malt (diastatic)	0.5	0.5	0.5	0.5	0.5	0.5	0.5	0.5	0.5
Dough									
Flour—clear	5	5	5	—	—	10	—	—	—
—patent	—	—	—	—	—	—	—	15	5
—light rye	25	17	—	24	24	30	20	15	—
—dark rye	—	8	25	10	10	—	10	10	35
Water[c]	59.5	61.5	64	57	62	62	64	64	64
Yeast	—	—	—	—	—	0.25	1	0.25	—
Salt	2	2.25	2.5	2	2	2	2.25	1.75	5–7
Shortening	3	3	3	4	2	1	—	2	4
Molasses	2	2	2	—	2	—	—	—	—
Sugar	—	—	—	—	—	—	—	4	—
Caraway seeds	1	1	1	—	1.75	—	—	—	3
Raisins	—	—	—	—	—	—	—	35	—
Cheese (grated)	—	—	—	—	—	—	20	—	—
Syrup	—	—	—	12	—	—	—	—	—
Nonfat dry milk	—	—	—	4	—	—	—	—	—

[a]All ingredients are listed as percent of flour weight. Flour weight is the total of the wheat flour plus various rye flours.
[b]% based on sponge flour.
[c]% based on sponge and dough flour.
Source: From Richards (1937), Slaten (1960), and Lorenz (1987).

Table 5 Rye Bread Formulations (Using a Sour Dough), %[a]

Ingredient	Heavy sour rye	German sour rye	Milwaukee sour rye	California sour rye
Flour, Clear	100	80	67	65
Light rye	—	10	—	—
Medium rye	—	—	17	23
Dark rye	—	10	16	12
Water	60	60	56	59–61
Yeast	0.25	1.75	0.50	1.0
Salt	2.5	2.0	2.0	2.5
Shortening	1.0	—	—	1.5
Caraway seeds	1.0	1.5	0.75	1.0
Malt	—	—	—	1.5
Prepared culture	4–6	—	—	—
Rye sour	40–60	60	4.0	6.0

[a]All ingredients are listed as a percentage of flour weight. Flour weight is the total of the wheat flour plus various rye flours.

Source: Data from Richards (1937), Slaten (1960), Sultan (1965), and Lorenz (1987).

pH value can be reached by using a sour that provides the best growing conditions for hetero and homofermentative lactobacilli. Optimal pH values can also be reached by including in the bread formulation either acidophilus milk, cultured buttermilk, or plain buttermilk. Sourdough breads require a long production time for superior flavor, crumb characteristics, and shelf life.

Fermentation times vary depending on type of wheat and rye flours used and whether or not a sour is used (Pyler, 1973; Lorenz, 1987).

C. Makeup

Properly fermented rye doughs should not cause any unusual problems during mechanical makeup, especially if the doughs contain less than 50% of medium rye flour. Standard dividers and rounders handle rye doughs as readily as they do wheat flour doughs.

Sponge doughs, being more pliable, drier, and less gassy than straight doughs, will adapt themselves particularly well to divider and rounder operations. Thus, they pass through the divider with less punishment and more accurate scaling, and through the rounder with a minimal need for dusting flour.

Rye breads are baked in pans, on screens, or as free standing loaves (hearth).

D. Proofing

The final proof of the molded pieces should be carried out at a temperature of 92–95 °F (33–35 °C) and at a relative humidity of 85%. In general, an average proofing time of 35–45 min should suffice for most conditions. The correct degree of proof is largely governed by the previous and subsequent handling of the dough and must be determined by practical experimentation and observation. Very general rules applying to rye dough proofing include the following: The darker the rye doughs, the less proof they require; also, the greater the percentage of rye flour in the dough, the shorter the proof should be. Doughs containing 2% yeast require a shorter proof time than when lower yeast additions are employed (Meyer, 1965; Pyler, 1973; Lorenz, 1987).

E. Washing/Glazing

To produce a healthy, brilliant crust bloom, the loaves are washed before the cutting or docking operations. Washing is of no value if the loaves are allowed to dry before being placed in the oven. The "wash" may be plain water, egg wash (whole eggs water 50:50), or a thin solution of gelatinized starch (20 g corn or wheat starch in 1 L boiling water or 3.9 oz per 10 lb of water). The latter produces a high gloss and is more effective if applied when the loaves come from the oven. When the starch wash is used after baking, it should be applied to the loaves while they are still hot. If the loaves are allowed to cool before the starch wash is applied, they will not have the desired gloss. With the proper steam condition in the oven, it is usually not necessary to wash the loaves before placing them in the oven unless they have become crusted because of dry proofing conditions (Meyer, 1965; Pyler, 1973).

F. Cutting and Docking

The heavy types of rye bread should be docked with a sharp stick or spike about the size of an ice pick. The lighter types of rye bread should be cut straight across the top of the loaf. A little experience and judgment is required to determine the correct depth of the cuts. In loaves that are given a relatively short proof, the docking and cutting should be deeper than for a longer proof. The cutting and docking of rye bread should be done immediately before placing the loaves in the oven (Lorenz, 1987).

G. Baking

Bread quality, and especially bread flavor, are influenced by the baking process. Optimal baking temperature and time depend on the weight of the dough pieces and the shape of the loaf. Both parameters determine the rates of heat transfer into the center of the loaves.

Rye bread in general requires a stable, solid heat within a temperature range of 470°–500 °F (245°–260 °C). Rye doughs from which milk and sugar are omitted require the higher temperature, as do doughs containing the lighter rye flours. Darker rye doughs have a tendency to color more rapidly because of their greater enzyme activity and are properly baked at the lower temperature. The oven heat must be so adjusted to produce a uniform crust on both bottom and top of the loaf. Drastic differences between top and bottom heat must be avoided. Excessive bottom heat results in too rapid an oven spring, which causes unsightly cracks and round bottoms. Too low bottom heat fails to produce an adequate expansion of the loaf, which assumes a flat form.

Low-pressure steam should be injected into the oven a few minutes before it is loaded and turned off a few minutes after loading is completed. In loading the oven, care should be taken not to place free-standing loaves too close together. If the loaves are too close together in the oven, breaking and splitting of the crust will occur (Slaten, 1960; Doose, 1964; Meyer, 1965; Pyler, 1973).

X. RYE BREAD FAULTS

Loaf symmetry is very important because of its effect on slicing properties, as most rye bread produced in large plants reaches the consumer sliced and wrapped.

There are several factors that will cause lack of symmetry. Carelessness in the makeup process, whether by machine or hand, is one of the greatest causes of rye bread cripples. A molder that is not adjusted properly can result in many cripples. The correct control of oven conditions is of the utmost importance. Oven heat must be solid and not flashy, with both bottom and top heat so regulated as to produce a uniform crust on both bottom and top of the finished loaves. This means that if the oven has excessive bottom heat, the bread will spring too rapidly, forming wild breaks and round bottoms. On the other hand, insufficient bottom heat results in flat loaves. An oven with a low crown gives best results because the steam injected into this type oven is forced down over the loaves. This helps to produce symmetrical loaves and also prevents cracking during baking. The use of low-pressure steam is essential in this connection, and the steam should be injected into the oven a few minutes before loading and turned off a few minutes after loading is completed.

Proofing is also very important—a high relative humidity should be maintained in the proof box to prevent crusting. This crusting will cause the loaves to crack and burst in the oven—a difficulty that holds true also when insufficient proof is given (Meyer, 1965). Rye bread will invariably crack or burst during baking if doughs are taken on the young side. Avoid underfermented doughs. Rye bread faults and their causes are summarized in Table 6.

A defect that is occasionally observed in rye bread is the so-called water ring, which assumes the form of a circular crumb zone about an inch beneath the crust and possessing a firmer, denser, and somewhat darker cell structure.

Its formation is due to a collapse of the cell structure as a result of crumb shrinkage during bread cooling. The defect is observed only in hearth rye breads made with wheat flour admixtures, and not in rye breads made from all-rye flour. Preventive measures include adequate acidification of the dough, use of strong clear wheat flours, correct dough fermentation, and more thorough baking to strengthen the crumb structure (Pyler, 1973). Rye bread quality faults traceable to the sour are summarized in Table 7.

XI. SLICING AND PACKAGING

Most rye bread produced in wholesale plants reaches the consumer sliced and wrapped. The need for slicing rye bread accounts for the fact that the baking of rye bread directly on the hearth is no longer as prevalent as it was formerly. Hearth-baked loaves lack uniformity of length, which causes difficulties in slicing. This problem has been eliminated by the use of so-called basket pans, which produce loaves of uniform length and with blunt ends (Pyler, 1973).

Practical experience has confirmed the greater suitability of the reciprocal slicer for rye bread slicing as opposed to the band slicer. The reciprocal slicer, because of its alternating cutting action, is less subject to the gumming up of blades than is the band slicer in which the blades travel

Table 6 Rye Bread Faults

1) **Lack of Volume**
   - —overmixed dough
   - —excessive pan proof
   - —young or old dough
   - —poor-quality flour

2) **Crust**
   - a) **pale**
     - —old dough
     - —lack of oven steam
     - —low baking temperature
     - —baking time too short
   - b) **dull or streaky**
     - —lack of steam during baking
     - —steam too dry
     - —excess dusting flour
     - —low baking temperature
     - —overaged dough
     - —cool oven
   - c) **blistered**
     - —slack young dough
     - —excessive oven steam
     - —hot wet proofing
     - —excessive top heat during baking
   - d) **thick**
     - —hot dry proof
     - —old dough
     - —lack of steam
     - —cool oven
     - —overbaking

3) **Symmetry of Form**
   - a) **burst sides**
     - —hot oven
     - —improper panning—panned with seam on top
     - —improper spacing of pans or bread in oven
     - —overmixing
     - —overaged dough
   - b) **caved-in sides**
     - —weak flour
     - —old dough
     - —cool oven
     - —underbaking
     - —overproofing
   - c) **flat top**
     - —overmixing
     - —slack dough
     - —young dough
     - —excessive moisture in proof box

4) **Grain**
   - a) **coarse**
     - —young or old dough
     - —under or overmixing
     - —slack dough
     - —cool oven
   - b) **open**
     - —slack dough
     - —excessive pan proof
     - —overfermentation
     - —loose molding
     - —cool oven

5) **Crumb color—gray streaky**
   - —excessive dusting flour
   - —overfermented dough
   - —immature dough
   - —coarse, open grain

6) **Texture—harsh or coarse**
   - —insufficient mixing
   - —old or young dough
   - —excessive baking
   - —dull slicing blades
   - —overproofing
   - —cool oven

7) **Flavor and Taste**
   - a) **lack of**
     - —young dough
     - —insufficient salt
     - —type of flour
     - —overproofing
   - b) **strong, objectionable**
     - —poor quality of ingredients
     - —overfermented dough
     - —underbaking
     - —improper flour storage
     - —rancidity of an ingredient
     - —wrapping product too warm

Source: Meyer (1965), Pyler (1973), and Lorenz (1987).

Table 7 Rye Bread Faults Traceable to Sour

Sour	Faults	Remedy
Too young	Small volume, cracks in crust, very compact, uneven grain, crumbly cracks in crumb, uneven crumb color, bland taste, unsatisfactory crumb elasticity.	Longer ripening time for sour, slightly higher temperatures, larger percentage of flour in sour, higher flour percentage in basic sour.
Too old	Small volume, flat loaves, sour taste, very uneven grain, gummy streaks.	Start with new pure culture sour or prepare new starter as described above.
Not enough sour	Very fine, compact grain, crumbly weak crumb elasticity, bland taste.	Increase percentage of flour in sour
Too much sour	Flat loaves, sour taste, uneven grain, gummy streaks.	Reduce sour to level appropriate for type of bread.
Too warm	Uneven, but compact grain, cracks, sour taste, gummy streaks.	Adjust sour temperature.
Too cold	Small volume, compact grain, sour taste, reduced crumb elasticity.	Adjust sour temperature.

Source: Lorenz (1983, 1987).

continuously in one direction. However, the use of the band slicer is dictated in large volume production by economic considerations. The use of shortening, as has been pointed out previously, greatly reduces the tendency of rye breads to gum up slicer blades. Slicing is further facilitated by the proper cooling of rye breads, which should not reach the slicing machine until the interior loaf temperature has reached 90 °F (32 °C) or less. As a rule, rye bread requires a longer cooling period than white bread under otherwise identical conditions (Meyer, 1965; Pyler, 1973).

Packaged sliced rye bread has to be protected against molds by using preservative agents in the formulation. The use of sourdough in rye bread production postpones mold formation by 1–2 days, but the reason for this phenomenon has so far not been fully explained (Seibel and Brümmer, 1991).

XII. PUMPERNICKEL

Pumpernickel was introduced in Germany around 1540 in the area surrounding the town of Osnabrück in Westphalia (Doose, 1948). This type of bread was first baked during a time of famine. Bread was about all there was to eat for a large segment of the population. Therefore, bread had to be flavorful, and Pumpernickel has a very distinct and characteristic flavor. Pumpernickel became very popular after World War II and is now produced in many countries where rye is grown or where rye flours are available to the baker.

Many of the breads called Pumpernickel in the United States are not real Pumpernickel. It takes a relatively long time to bake real Pumpernickel as well as a special flour and special equipment, which often are not available. Therefore, real Pumpernickel is produced by only a few bakeries in the United States.

A typical Pumpernickel formulation and procedure are given in Table 8. Some bakers use a preferment in the production of Pumpernickel (Brühstück). This involves soaking up to 10% of the whole rye meal in water (3 parts of water to 1 part of rye meal). The temperature of the water added ranges from 70 °C to 100 °C, resulting in a preferment temperature of about 50 °C. This soaking produces thorough hydration of the rye meal, partial gelatinization of the starch (gelatinization of rye starch occurs at about 56 °C), and some enzyme hydrolysis to produce sugars that will influence the flavor of the final product. Using more than 10% of the rye meal in the formulation for the Brühstück results in detrimental effects on crumb elasticity.

The addition of residue Pumpernickel—returns, damaged loaves, crumbs and bread pieces recovered during slicing—increases water binding capacity, improves dough handling, provides additional color, contributes to the desirable spicy flavor of the bread, and extends shelf life.

Roasted residue Pumpernickel is soaked overnight in warm water, then cooled before addition to the other dough ingredients. During the 8- to 12-h soaking period, organic acids are formed along with other flavor components that contribute greatly to the flavor characteristics of the final product.

There is a limit, however, to the amount of residue bread that can be added. Amounts over 10% of total rye meal weight will reduce crumb elasticity and may cause a gummy crumb and a very compact grain. Bread crumb characteristics are not affected using 10% residue bread if the bread is baked between 120 °C and 170 °C. Baking at 100 °C requires a reduction of residue bread to 7% of total rye meal or less (Lorenz, 1980).

Table 8 Typical Pumpernickel Formula and Procedure

Sour	**Grams**	**Pounds**
Basic sour (multiple stage, Table 3)	500	1.0
Whole rye meal	1000	2.0
Water	750	1.5
Temp.: 82 °F (28 °C)		
Ripening time: 2–3 h		
Residue Bread[a]	**Grams**	**Pounds**
Ground, roasted pumpernickel[b]	1500	3.0
Water [158–212 °F (70–100 °C)]	3000	6.0
Hold in Proof Box 8 h at 95 °F (35 °C)		
Dough	**Grams**	**Pounds**
Sour	2250	4.5
Residue bread	4500	9.0
Whole rye meal	9000	18.0
Water	2000	4.0
Salt	150	0.3
Compressed yeast	50	0.12

Procedure

1. Mixing time: 20–25 min at low speed.
2. Ferment 15 min, then remix 10 min. Dough temp.: 82 °F (28 °C).
3. Divide and shape loaves.
4. Proof 50–60 min.
5. Bake 20 h at 230 °F (118 °C) with steam in oven.
6. Cool in pans before depanning, slice after 1 or 2 days, unwrapped, on racks.

[a]Residue Bread—anyone or all of the below:
Loaves damaged during slicing and wrapping.
End pieces after slicing.
Crumbs from slicer.

[b]Ground roasted pumpernickel—ground, placed on trays and placed into oven. Length of time and oven temperature variable depending on extent of roasting desired.

Source: Doose (1948) and Lorenz (1980).

Pumpernickel is baked in fully closed pans in a special baking chamber. The pans are not only placed next to each other in the oven but also on top of each other. The steam introduced into the chamber is absolutely essential for optimal product quality. The baking time is up to 24 h depending on the baking temperature, which is in the range of 100–170 °C. A baking time of 16 h is considered minimal. For most desirable product color, baking times of 20–24 h are recommended.

After baking, the covers are removed to allow the steam to escape and the bread is permitted to "sweat" in the pans for about 15 min. The formation of moisture between crust and the walls of the pans makes it easier to remove the loaves from the pans.

The sweet taste of Pumpernickel is due not to the addition of sweeteners to the formulation but rather to enzymatic hydrolysis of the starch during sour ripening, dough preparation, and proofing as well as the special baking process. The low baking temperature of 100–170 °C for 16–24 h in a high-moisture environment and a rather low pH continues to modify the starch extensively, forming glucose, maltose, and dextrins. The internal temperature of the bread during baking does not reach the inactivation temperatures of the amylases for about 12 h. The sugar content of Pumpernickel has been found to increase for the first 12 h of baking and then level off. Final sugar contents range from 11% to 26% compared to about 7% for regular rye bread baked without sweeteners (Doose, 1948). Caramelization of the sugars and the development of Maillard reaction end products greatly affect the flavor and taste of Pumpernickel.

Depending on formulation and processing procedure, the following four basic Pumpernickel flavor types can be distinguished.

Flavor type	*Method of preparation*
1. Sweet	0.4% yeast, no sour
2. Sweet–slightly sour	0.3% yeast, 7–10% of rye meal in sour
3. Sweet–sour 4. Sour–sweet	No yeast, 10–30% of rye meal in sour

XIII. PUMPERNICKEL BREAD FAULTS

There are relatively few possible Pumpernickel bread faults that can be identified, but they are different from those seen in other rye breads and are due to the special manufacturing process (Lorenz, 1980; Doose, 1948).

The following bread faults have been identified:

Wet product
Shrunken sides
Uneven grain
Weak crumb elasticity
Shrunken crumb

Preparation:

Disperse Saf Levain in water at a temperature of 30–35 °C.
Add the flour and salt.
Mix on slow speed, ensuring that the dough temperature does not exceed 27 °C (8 min with a wishbone mixer, 5 min on spiral mixer, and 5 min on a planetary mixer).
Fermentation time: 24 h at 30 °C.

Examples of applications in breadmaking. See Tables 12–15.

C. Paste or Liquid Starters

These are sourdough concentrates with a short life (less than 3 weeks), for making one-stage full sourdoughs. They must be stored at 4 °C. The quantity used is about 5–10% of the weight of flour used in the sourdough. They give the same results as those obtained with powdered starters but their shelf life restricts their use. The main products on the French market are from Böcker (made in Germany), Philibert, and Puratos.

III. READY-TO-USE SOURDOUGHS

Although mixed starters have simplified sourdough production, the baker is still reliant on preparing full sourdough, which takes 16–24 h and requires rooms or tanks for controlling fermentation temperature, in particular. The ideal situation is therefore to have a sourdough available at any time that is simple and quick to use, so that the baker can react quickly to demand. This type of product is available from some manufacturers in Europe, in liquid form (Lesaffre, Philibert, Puratos). However, it is not easy to guarantee an active sourdough that loses none of its fermenting capacity from manufacture to the time of use. Few products on the market are stable enough over time; the viability of microorganisms drops very quickly after a few days' storage, even at low temperatures. One of the reasons for this instability is the excessive acidification that occurs when there is no refreshment. In a few days, even at 4 °C, this acidification results in the disappearance of most of the viable microorganisms in the sourdough, i.e., those that are capable of multiplying or fermenting the dough of a sourdough bread. The freezing or drying of firm or liquid sourdoughs is the physical treatment that disrupts the membranes of the microorganisms, preventing them from resuming their metabolic activity when they are used again (thawing or rehydration). Lesaffre is the only manufacturer in France in 2002 guaranteeing a storage time of 6 weeks in active form for its ready-

Table 12 Farmhouse Bread (250 g baked weight)

Formula

Ingredients	Weight (kg)	%
Wheat flour T65	9	90
Rye flour T170	1	10
Water	5.500	55
Salt	0.210	2.1
Yeast	0.220	2.2
Liquid sourdough (Saf Levain LV1)	3	30
Dough improver	+	
Total dough	18.930	

Process

	Mixers		
Mixing	Wishbone[a]	Spiral[b]	Planetary
1st speed	5 min	3 min	5 min
2nd speed	10 min	4 min	5 min

Mixing time given for:
[a]Wishbone: 1st speed 60 rpm/2nd speed 90 rpm
[b]Spiral: 1st speed 110 rpm/2nd speed 220 rpm

Add the sourdough at the beginning of the 2nd speed
Temperature at end of mixing: 25 °C
Floor time: 30 min
Dividing: 350-g dough pieces (slight rounding advisable)
Intermediate proof: 20 min
Molding: manual or mechanical
Proofing: 1 h 30 at room temperature
Baking: deck oven 35 min at 240 °C

Table 13 Traditional Baguette (250 g baked weight)

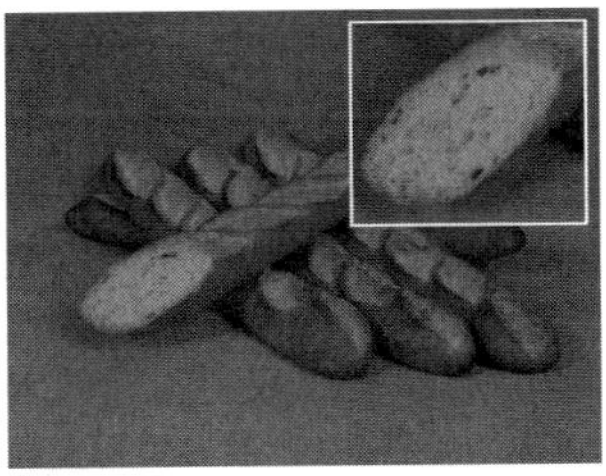

Formula

Ingredients	Weight (kg)	%
Traditional wheat flour (no additives)	10	100
Water	6.250	62.5
Salt	0.200	2
Yeast	0.200	2
Stiff sourdough (Saf Levain LV1)	2	20
Total dough	18.650	

Process

	Mixers		
Mixing	Wish bone[a]	Spiral[b]	Planetary
1st speed	5 min	3 min	5 min
2nd speed	9 min	3.5 min	6 min

Mixing time given for:
[a]Wishbone: 1st speed 60 rpm/2nd speed 90 rpm
[b]Spiral: 1st speed 110 rpm/2nd speed 220 rpm

Add the sourdough at the beginning of the 2nd speed
Temperature at end of mixing: 25 °C
Floor time: 45 min, knocked back halfway through
Dividing: 350-g dough pieces
Intermediate proof: 20 min
Molding: manual or mechanical
Proof time: 1 h 30 at 27 °C
Baking: deck oven 25 min at 240 °C

Table 14 Pain au Levain (Slow Proof): Traditional French Sourdough Bread According to French Legislation (250 g baked weight)

Formula

Ingredients	Weight(kg)	%
Wheat flour T65	10	100
Water	6.600	66
Salt	0.210	2.1
Yeast	0.020	0.2
Stiff sourdough (Saf Levain LV2)	4	40
Total dough	20.830	

Process

	Mixers		
Mixing	Wish bone[a]	Spiral[b]	Planetary
1st speed	8 min	4 min	5 min
2nd speed	4 min	1.5 min	3.5 min

Mixing time given for:
[a]Wishbone: 1st speed 60 rpm/2nd speed 90 rpm
[b]Spiral: 1st speed 110 rpm/2nd speed 220 rpm

Add the sourdough at the beginning of the 2nd speed
Temperature at end of mixing: 26 °C
Floor time: 1 h 30 knock back halfway through
Dividing: 350-g dough pieces (slight rounding advisable)
Intermediate proof: 25 min
Molding: manual
Proof time: 14 h at 14 °C
Baking: deck oven 30 min at 235 °C

Table 15 San Francisco-type Sourdough Bread (retarded dough; 250 g baked weight)

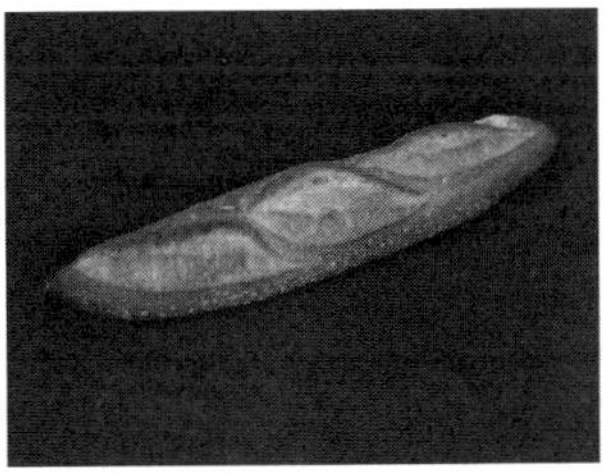

Formula

Ingredients	Weight (kg)	%
Wheat flour T65	10	100
Water	6.400	64
Salt	0.200	2
Stiff sourdough (Saf Levain LV3)	3	30
Yeast	0.050	0.5
Total dough	19.650	

Process

	Mixers		
Mixing	Wish bone[a]	Spiral[b]	Planetary
1st speed	12 min	8 min	8 min
2nd speed	2 min	1 min	2 min

Mixing time given for:
[a]Wishbone: 1st speed 60 rpm/2nd speed 90 rpm
[b]Spiral: 1st speed 110 rpm/2nd speed 220 rpm

Add the sourdough at the beginning of mixing
Temperature at end of mixing: 24 °C
Floor time: 1 h
Punch down and retard in cold room at 4 °C
Warm up at room temperature (core temperature must rise up to 12 °C)
Dividing: 350-g dough pieces (slight rounding advisable)
Intermediate proof: 20 min
Molding: manual
Proof time: 1 h at room temperature
Baking: deck oven 30 min at 235 °C (declining temperature)

to-use liquid sourdough, at a temperature of 0–6 °C, maximum, thanks to a patented production process (European Patent Application EP 0 953 288 A1, product marketed under the brand name of Crème de Levain).

A. Crème de Levain Manufacturing Process

1. Aim

To preserve the sourdough, i.e., preserve a living biomass of lactic acid bacteria and sourdough yeasts.

The quantity and physiological condition of these microorganisms must be such that, after storage, bread with the characteristics of a loaf made with a spontaneous sourdough can be fermented and produced.

2. Methods

Research has shown that the storage of sourdough depends on its pH, the quantities of acetic and lactic acid, and the quantity of ethanol and directly fermentable sugars present in the sourdough.

Only the nondissociated forms of organic acids are toxic to microorganisms as they can cross the cell membrane. This dissociation depends on the pH of the medium; the higher the pH, the more dissociated the organic acid, which helps to preserve the sourdough's properties. Acetic acid is by far the most toxic since at pH 4.3 it is mainly in nondissociated form whereas lactic acid is almost wholly nondissociated.

In the process, the relative proportions of acetic and lactic acid are controlled by checking the available fermentable sugars. This mainly depends on the quantity of fermentable sugars that have been added during the multiplication phase of the lactic acid bacteria (sourdough start), and on the quantity of starch that can be degraded into maltose and glucose, from the flours used. The starch is hydrolyzed by excess quantities of amylases, a process that releases all the fermentable sugars from flour as the sourdough is prepared. When the latter has matured and reached the required pH, and as soon as the substrate has been exhausted by the microorganisms, there is only residual amylase activity and the sourdough can commence storage.

As mentioned previously, ethanol production must be controlled too. Apart from the limitation regarding the quantity of substrate, choosing a strain of yeast that does not ferment maltose restricts how much is produced. This characteristic favors the metabolism of lactobacilli for which maltose is the main available fermentable sugar.

To prevent the sourdough settling, it is stabilized with a traditional food stabilizer, such as xantham gum, or stored while being slowly stirred.

B. Applications in Breadmaking

The examples shown in Tables 16–19 illustrate applications of Crème de Levain ready-to-use sourdough.

IV. CONCLUSION

Although the use of sourdough in breadmaking almost died out in France after the early 1950s, there is now a renewed interest in its use among bakeries. Those involved in the wheat, flour, and bread industries want to play the flavor card as a means of curbing the fall in bread consumption seen in recent decades, a move that has found an echo with manufacturers specializing in ingredients for bread fermentation. The baker now has products that allow him to make sourdough bread in a way that is in keeping with socio-economic reality in industrialized Western societies. It is also a way of giving consumers what they want because, in addition to being a means of making bread, sourdough is a natural flavor enhancer that can be used in all types of products—croissants, brioches, pizzas—or in rapid breadmaking processes. It is also a "natural" preservative that precludes or limits the use of additives.

ACKNOWLEDGMENTS

Thanks to Pascal Lejeune, sourdough specialist in the research and development department of Lesaffre International, for the accuracy of his information, the relevance of his comments, and his constructive criticism. Thanks also to Isabelle Debourse for her help with laying out the data. Finally, thanks to Dorothy Ellerington for her remarkable ability to translate from French to English in the context of a very unusual topic.

Table 16 Old Style Baguette (250 g baked weight)

Formula

Ingredients	Weight (kg)	%
Wheat flour T65	10	100
Ready-to-use sourdough (Crème de Levain) at 4 °C	1	10
Water	5.800	58
Salt	0.200	2
Yeast	0.150	1.5
Dough conditioner	+	
Total dough	17.150	

Process

	Mixers		
Mixing	Wish bone[a]	Spiral[b]	Planetary
1st speed	10 min	6 min	6 min
2nd speed	6 min	2 min	4 min

Mixing time given for:
[a]Wishbone: 1st speed 60 rpm/2nd speed 90 rpm
[b]Spiral: 1st speed 110 rpm/2nd speed 220 rpm

Add the sourdough with water
Temperature at end of mixing: 24 °C
Floor time: 1 h 30, knock back halfway through
Dividing: 350-g dough pieces
Intermediate proof: 20 min
Molding: manual or mechanical (sheeter rolls wide open)
Proof time: 1 h at 25 °C
Baking: deck oven 25 min at 230 °C

Table 17 Traditional Baguette (250 g baked weight)

Formula

Ingredients	Weight (kg)	%
Wheat flour T65	10	100
Ready-to-use sourdough (Crème de Levain) at 4 °C	1.500	15
Water	5.000	50
Salt	0.200	2
Yeast	0.200	2
Total dough	16.900	

Process

	Mixers		
Mixing	Wishbone[a]	Spiral[b]	Planetary
1st speed	5 min	3 min	5 min
2nd speed	9 min	3.5 min	6 min

Mixing time given for:
[a]Wishbone: 1st speed 60 rpm/2nd speed 90 rpm
[b]Spiral: 1st speed 110 rpm/2nd speed 220 rpm

Add the sourdough with water
Temperature at end of mixing: 25 °C
Floor time: 45 min, knocked back halfway through
Dividing: 350-g dough pieces
Intermediate proof: 20 min
Molding: manual or mechanical
Proof time: 1 h 30 at 27 °C
Baking: deck oven, 25 min at 230 °C/240 °C; rack oven, 22 min at 230 °C/240 °C

Table 18 Sourdough-Type Bread (400 g baked weight)

Formula

Ingredients	Weight (kg)	%
Wheat flour T65	10	100
Ready-to-use sourdough (Crème de Levain) at 4 °C	1.500	15
Water	5.200	52
Salt	0.200	2
Yeast	0.150	1.5
Dough conditioner	+	
Total dough	17.050	

Process

	Mixers		
Mixing	Wishbone[a]	Spiral[b]	Planetary
1st speed	12 min	8 min	8 min
2nd speed	5 min	2 min	4 min

Mixing time given for:
[a]Wishbone: 1st speed 60 rpm/2nd speed 90 rpm
[b]Spiral: 1st speed 110 rpm/2nd speed 220 rpm

Add the sourdough with water
Temperature at end of mixing: 27 °C
Floor time: 1 h
Dividing: 500-g dough pieces (slight rounding advisable)
Intermediate proof: 30 min
Molding: manual or mechanical (sheeter rolls wide open)
Proof time: 1 h at 25 °C
Baking: deck oven, 40 min at 230 °C

Table 19 Brioches

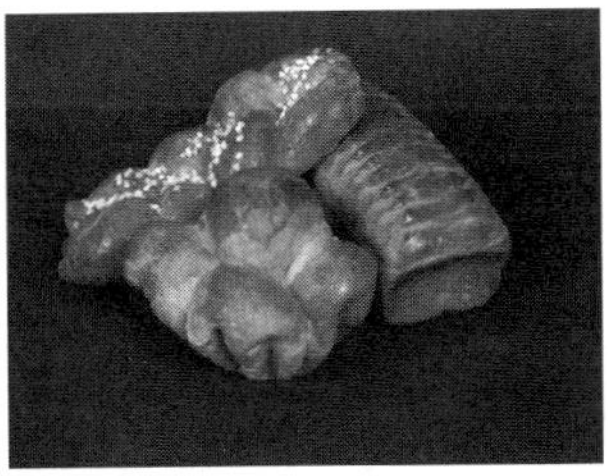

Formula

Ingredients	Weight (kg)	%
Wheat flour T45	10	100
Water or milk	1.800	18
Ready-to-use sourdough (Crème de Levain) at 4 °C	1.200	12
Yeast	0.500	5
Salt	0.200	2.0
Dough conditioner	+	
Butter (delayed)	2	20
Sugar (sucrose)	1.200	12
Eggs	3	30
Total dough	19.900	

Process

Mixing (spiral): 1st speed 5 min, 2nd speed 10 min—add butter and mix until clean-up
Temperature at end of mixing: 26 °C
Floor time: 45 min (room temperature) + 2 h (cold room)
Dividing: weight and number of pieces according to shape
Rounding
Intermediate proof: 25 min
Molding: manual or mechanical
Proof time: 1 h 45 at 28 °C
Baking at 180 °C

REFERENCES

1. H. Chiron and B. Godon. La panification française: historique de la panification (R. Guinet and B. Godon, eds.). Tec et Doc Lavoisier, Paris, 1994, pp. 5–44.
2. J. A. Barnett. Yeast—A History of Research on Yeasts: Louis Pasteur and His Contemporaries, 1850–1880, Vol. 16. John Wiley & Sons. London, 2000, pp. 755–771.
3. R. Calvel. Le pain et la panification. Que sais-je. No. 1140, PUF, Paris, 1964, pp. 84–88.
4. R. Calvel. Bulletin des anciens élèves et amis du professeur Calvel. Ma philosophie du pain, les caractéristiques du bon pain français. 1987, p. 1.
5. T. Thorgevsky. Ça m'intéresse. Pourquoi le pain est-il si mauvais? 20–26 (November 1984).
6. H. Chiron. Industrie des céréales. Incidence des mutations technologiques sur les caractéristiques du pain français. 87:5–12 (1994).
7. R. Guinet. Le pain français. Qualité d'un bon pain de France, 20:402 (1964).
8. Ph. Roussel et H. Chiron. Les pains français—évolution, qualité, production, Maé-Erti, Vesoul, 2002, p. 21.
9. B. Poitrenaud. Filière Gourmande. La clé du beau et du bon pain. 86 87 (January 1999).
10. Philippe Millet. Arômes et Ingrédients. A la recherche du goût perdu (April 1999).
11. B. Onno et Ph Roussel. Bactéries lactiques, Technologie et microbiologie de la panification au levain, Lorica, Uriage (1994).
12. J. Durand. Libération. Le pain Poilâne a perdu son mitron (November 2002).
13. B. Onno and L. Ragot. Industries des céréales. Elaboration d'un levain naturel. Aspects microbiologiques. 54:17 (1988).
14. M. Infantes et J. L. Schmidt. Sciences des aliments. Characterisation of the yeast flora of natural sourdoughs located in various French areas. 12:271–287 (1992).
15. M. Infantes et C. Tourneur. Sciences des aliments. Survey on the lactic flora of natural sourdoughs located in various French areas. 11:527–545 (1991).
16. R. Calvel. Le Boulanger-Pâtissier. Fermentation et panification au levain naturel. 5, 6, 7, 8, 10 (1980).
17. B. Onno. La panification française. Les levains (R. Guinet and B. Godon, eds.). Tec et Doc Lavoisier, Paris, 1994, pp. 201–215.

9
Commercial Starters in Spain

Cristina M. Rosell and Carmen Benedito
Instituto de Agroquímica y Tecnología de Alimentos, Valencia, Spain

I. INTRODUCTION

The use of sourdough in breadmaking is an ancient practice, common in Spanish bakeries and in other European countries (Barber et al., 1981). Traditionally, many positive characteristics on the quality of both the bread dough and bread have been attributed to the sourdough. In the bakery, it is considered that the benefits of sourdough utilization are as follows:

reduces the development time of the dough
shortens the period of fermentation
favors volume development in the oven
improves the volume, aroma, and flavor of the breads

II. COMPOSITION OF SPANISH SOURDOUGH

Sourdough provides microorganisms needed for fermentation, contributing CO_2 evolution, rheological changes, and flavor development (Barber et al., 1981, 1983, 1991a,b; Hansen and Hansen, 1996; Wehrle and Arendt, 1998).

The microbial flora is a biological system, alive and dynamic, that requires precise and controlled conditions for its correct evolution and adequate activity. The initial microbial flora of the sourdough comes from the raw materials used in its formulation and from the ambient conditions. The conditions under which the fermentative process is developed determine the selective growth of the different microorganisms present in the dough and, because of that, the composition and activity of the final microbial flora. This is mainly composed of yeasts and lactic acid bacteria, the lactobacilli being predominant in this group (Barber et al., 1983; Barber and Baguena, 1988; Oura et al., 1982). Table 1 shows the composition of different sourdoughs taken from bakeries (Barber et al., 1983).

The presence of a complex microbial flora leads to simultaneous production of several types of fermentations (alcoholic, lactic, acetic, and others). Only the appropriate ratio of these fermentations produces sourdoughs with the desired functional properties. For that, knowledge of the microorganisms present in a sourdough (type, specie, and relative proportion) and of the conditions necessary to favor the appropriate development of the groups and species of more interest for fermentative process is needed.

Several studies have been developed to characterize the microbial flora of the Spanish sourdough (Barber et al., 1981, 1983, 1985a–c, 1987; Barber y Baguena, 1988). From them, it can be concluded that sourdough was composed of:

1. Yeasts, i.e., *Saccharomyces cerevisiae*
2. Lactobacilli:
    - *Lactobacillus brevis*
    - *Lactobacillus plantarum* (subspecies: *plantarum* and *arabinosus*)

Also found were other yeasts and other lactobacilli but in minor proportions.

The fermentative activity of the sourdough (changes in pH, total titratable acidity, production of acids) is dependent on the species,

Table 1 Microbial Viable Counts (cfu/g) in Different Industrial Sourdoughs from Wheat Flour

	Microbial viable counts (cfu/g)		
Sourdough	SD-1	SD-2	SD-3
Yeasts	1.28×10^8	2.25×10^4	1.10×10^7
Lactic acid bacteria	1.88×10^5	1.49×10^8	8.33×10^6

Source: Data from Barber et al. (1983).

subspecies, and even strain (Spicher, 1982). This activity is also influenced by the medium conditions (substrate composition, dough consistency, aeration, temperature, time, and pH) and the conditions for their performance (Spicher and Stephan, 1982).

III. PERFORMANCE OF THE SPANISH SOURDOUGHS

Diversification in formulation and preparative methods of the Spanish sourdoughs commonly employed have been reviewed by Martínez-Anaya (1983) and Barber (1992). The sourdoughs used in Spain are called mother doughs, but these are not as acidic as those used in another countries. These sourdoughs can be classified as follows:

1. Bread dough, obtained from a portion of dough of a previous breadmaking process. The so obtained dough, adequately fermented, is used to formulate a new dough to prepare bread or "is refreshed" by mixing it again with several times its weight of flour and the appropriate quantity of water, and leaving it to ferment until the moment it is needed. This type is the most commonly used in Spain.
2. Sponge dough, prepared by mixing flour, water, salt, and compressed yeast (Fig. 1A).
3. Spontaneous sourdough, obtained by mixing flour, water, and salt and with successive refreshings (i.e., mixing several times with flour and an appropriate amount of water), as indicated in Fig. 1B. This type is not commonly used nowadays in Spain, probably because it requires expertise and is time consuming.

Previous studies (Barber et al., 1981, 1983, 1985a–c; Barber and Baguena, 1988) showed that differences in the formulation and process conditions of commercial Spanish sourdoughs lead to different microbiological composition (qualitative and quantitative), which no doubt directly affect their functionality in breadmaking. In fact, Barber et al. (1989a,b, 1991b) investigated the relationship between the propagation of the microbial flora during the performance of a sourdough by the method of successive refreshing, and the functional properties of the bread dough and fresh bread quality. They showed an increase of the microbial viable counts in the sourdough with the increase in numbers of processing stages (Table 2). Improvement in bread characteristics (volume, density, texture, and acidity) were also detected as the number of refreshing stages increased (Table 3).

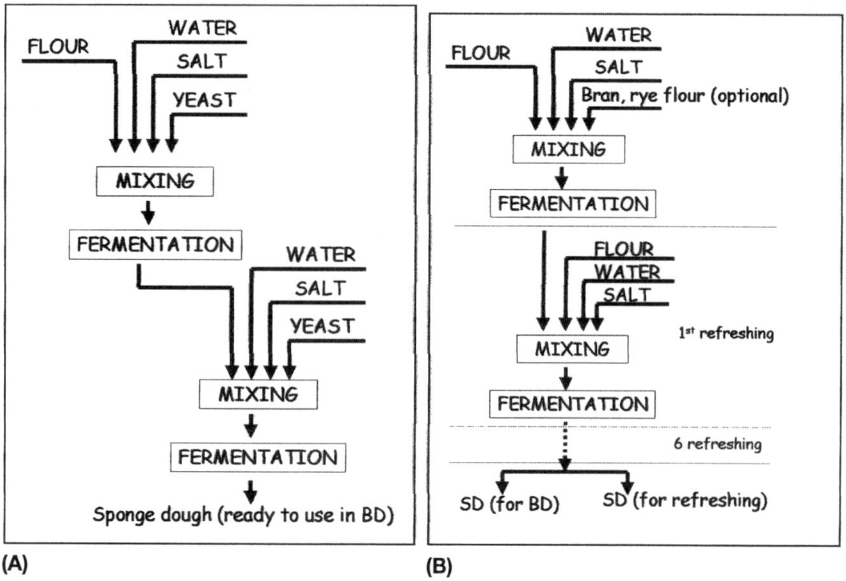

Figure 1 Sourdough preparation. (A) Sponge type. (B) Spontaneous sourdough.

IV. IMPORTANCE OF USING SOURDOUGHS IN BREADMAKING

Sourdoughs are a source of microorganisms and also provide different acids, which are products of several fermentations (alcoholic, heterolactic, acetic, and others) (Barber et al., 1985c; Martínez-Anaya et al., 1990). The

Table 2 Microbial Viable Counts (cfu/g) of a Sourdough at Different Stages of a Successive Refreshing Process[a]

	Microbial viable counts (cfu/g)		
Sourdough	SD-1	SD-3	SD-5
Yeasts	4.3×10^3	4.0×10^2	4.0×10^3
Lactic acid bacteria	3.0×10^3	9.0×10^6	7.0×10^7

[a] SD-1 sourdough before the first refreshing; SD-3 sourdough after two refreshing stages; SD-5 sourdough after four refreshing stages.
Source: Data from Barber et al. (1989).

Table 3 Characteristics of Bread Made with Sourdough from Different Stages of a Successive Refreshing Process[a]

	Bread stage				
Sourdough	B1	B2	B3	B4	B5
Volume (mL)	173	184	196	230	241
Density (g/mL)	0.30	0.28	0.26	0.25	0.22
Firmness (g)	1420	475	383	363	182

[a] B1 to B5 elaborated with sourdough from 1 to 5 refreshing stage.
Source: Data from Barber et al. (1991).

presence of these acids, mainly lactic and acetic, produce a modification in the characteristics of the bread dough that is transmitted to the fresh bread. The sourdough exhibits benefits both in processing (rheological and fermentative tough characteristics) and in bread quality.

A. Process Stages

The organic acids present in a sourdough modify, depending on type and concentration, the protein and rheological characteristics of the dough (Tsen, 1966; Tanaka et al., 1967; Galal et al., 1978; Martínez-Anaya et al., 1989; Barber et al., 1989a,b; Gobbetti et al., 1995). Therefore, it is evident that the procedure of sourdough preparation that directly determines its composition in acids would have an influence on its behavior in bread dough.

The addition of sourdough to bread dough promotes the following benefits (Barber et al., 1981, 1987, 1991a; Martínez-Anaya et al., 1990) to the process:

Reduction of the bread dough development time and stability, and an increase of the mixing tolerance index (determined by the Brabender farinograph).

Decrease of the dough extensibility and an increase of the resistance to extension, both parameters measured by the Brabender extensigraph.

Increase in CO_2 production, and a more elastic and stable dough during proofing; in addition, bread doughs have lower pHs and higher total titratable acidity than those obtained from straight dough processes.

B. Bread Level

In addition to modifying the bread dough during the process, the presence of different acids from the sourdough leads to different characteristics of the final breads. There are several studies describing the effects of the addition of sourdough on bread quality characteristics (Barber et al., 1989a,b, 1991b, 1992b; Barber, 1992a; Martínez-Anaya et al., 1990, 1991). These studies led to the conclusion that the addition of sourdough leads to:

- An improvement of the taste and flavor of breads
- An increase in volume and a decrease in bread density
- An improvement of the crumb texture, given better grain and crumb softness
- An extension of the shelflife of breads due to the antimicrobial effect of the acids generated by the microflora

Nevertheless, as mentioned previously, the sourdoughs are composed of a microbial flora that is very sensitive to any changes in the medium, which may be reflected in bread dough and subsequently in the bread. Besides that there are other potential drawbacks:

- A high variability depending on sourdough performance, substrates, and medium conditions (temperature, pH, time, etc.)
- A very time-consuming process, requiring a daily preparation and keeping of adequate conditions
- The need to train or hire experienced personnel to obtain sourdoughs with uniform characteristics

These potential disadvantages have encouraged a number of manufacturers of breadmaking additives to seek practical solutions. In the last decade, different preparation methods have been developed to obtain a good sourdough that guarantees uniformity in breadmaking and bread quality.

V. PRODUCTS FOR THE PREPARATION OF SOURDOUGHS PRESENT IN THE SPANISH MARKET

Four types of products can be found in the Spanish market:

1. *Starter*: Mixture of appropriate microorganism for the preparation of food-quality sourdough.
2. *Sourdough*: Biologically active sourdough.
3. *Dehydrated or lyophilized sourdough*: Imparts acidity to the dough and affords fermentation aromas without having fermentative activity.

4. *Sourdough substitutes*: Mixture of additives and organic acids added to flour to strengthen the dough (inactive biologically).

Table 4 is a list of products present in the Spanish market and their producers.

A. Starters

The starters are being introduced in the Spanish baking industry very slowly, even though they are products of good quality. In the Spanish market there are products from two origins: France and Denmark, the first being a mixture of lactobacilli and yeast and the second pure lactobacilli. There are pure strains that have been selected from the natural microflora of sourdoughs and stabilized. The live microorganisms contained in the product are claimed to be known and controlled, conferring consistent characteristics to any of the successive sourdoughs.

With both starters it is possible to prepare sourdough in dough or liquid form. The recommended use level is around 0.5% flour basis for an appropriate sourdough. The starters can be used to prepare a liquid sourdough in batches of small or large quantity or even to use a programable fermentor. In any case, the mixture of flour and water (1:1) with the starter (0.5% in a flour basis) and salt is matured at 27–30 °C during 16–24 h according to the desired degree of acidity (Fig. 2). After a fast cooling, the matured liquid sourdough can be stored at 4 °C for 3–5 days.

Table 4 Some of the Commercial Products Present in the Spanish Market

Type of product	Commercial name	Suppliers
Starter	Gamme Florapan	Christian Hansen
	Gamme Saf Levain	Lesaffre
Sourdough	Comabans	Indespan
	Gamme Phil Levain	Philxn
Dehydrated sourdough	Gamme Arome Levain	Lesaffre
	Gamme Saskia	Fher, Spain
	Masaferm	Gist Brocades
	Gamme Phil	Philxn
Sourdough substitutes	Traviatta	T-500 Puratos
	Panipan Masa	Lasem, Spain
	Supral Vit	Fher, Spain

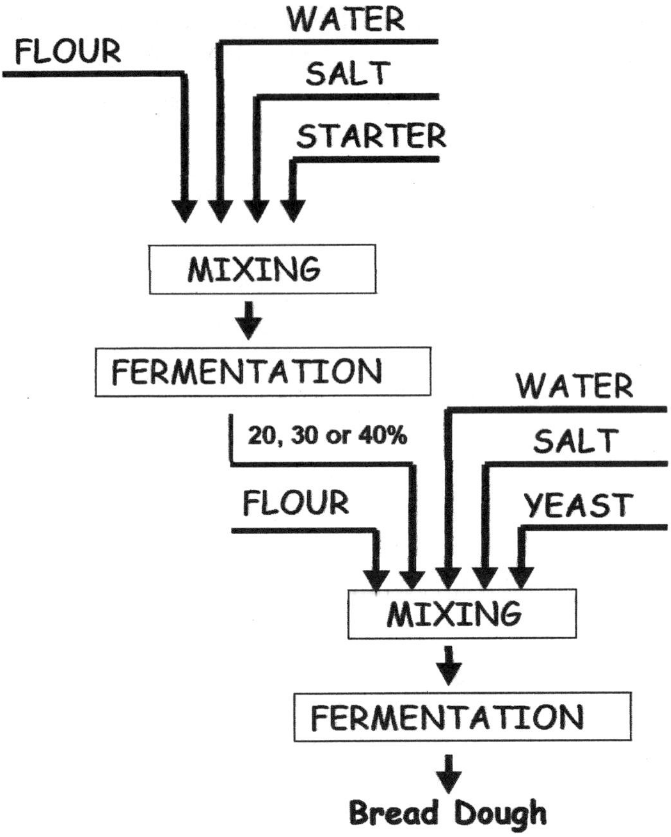

Figure 2 Use of starter to prepare a sourdough and a final bread dough.

The sourdough prepared in this way may be added in 20%, 30%, or 40% (flour basis) to bread dough.

Suppliers recommend to use the starter directly and not to submit it to refreshing. This practice avoids both the loss of viability of the lactobacilli in the sourdough and the risk of microbial contamination that will yield the growth of contaminating microorganisms, responsible for gluten degradation and unpleasant flavors.

The use of these products avoids the critical and uncertain phases of both preparation and refreshing of a sourdough by the traditional methods, and guarantees the uniformity of the sourdough. Suppliers claim that such sourdoughs presents the same aroma, taste, and rheological characteristics of a traditional sourdough obtained by spontaneous process, and that their use in breadmaking

Improves the taste and flavor of the bread
Improves crumb texture
Extends the bread shelf life
Confers to the bread the desirable acidity

These are marketed in powder form and should be stored at temperature between 0 °C and −25 °C. Nevertheless, they can be kept at ambient temperature for several hours, mainly during transportation.

In the authors' laboratory, numerous studies have been carried out focusing on the characterization of these starters, their manufacture, their use, the resulting sourdough properties, and the final effect on baking performance. The results obtained from those studies have been compiled in an extensive literature (Barber et al., 1983, 1985a–c; Barber and Baguena, 1989; Barber, 1992; Collar et al., 1991a–c, 1994, 1998; Martínez-Anaya et al., 1989, 1990, 1991, 1994a,b, 1998; Martínez-Anaya, 1994, 1996; Mascaros et al., 1994; Rouzaud and Martínez-Anaya, 1993, 1997; Torner et al., 1989, 1992).

B. Commercial Sourdoughs

The biologically active sourdoughs are the most recent substitutes for the sourdough present in the Spanish market. These products are commercialized in two forms: liquid and powdered.

1. Preparation

Biologically active sourdoughs are obtained by a spontaneous process using their own microbial flora called also mother sourdough, for instance, fermenting rye or wheat flour. The fermentation time and temperature vary according to the desired final characteristics of the product. The activity of the microbial flora is preserved by conditioning methods to ensure the uniformity and the quality of the products. In Table 5 are specified the characteristics of some sourdoughs given by the suppliers.

These sourdoughs differ in the final pH and total titratable acidity due to the different microbial flora, which is still viable, and the process conditions (substrate, temperature, time, etc.). In some cases, these sourdoughs are composed of the microbial flora and some bread improvers such as ascorbic acid, malted flour, and some enzymes commonly used in breadmaking.

Table 5 Characteristics of Some Commercial Sourdoughs from Wheat

Sample	Dosage (% f.b.)	pH (sol. 10%)	TTA (mL NaOH 0.1 N)	Yeasts (cfu/g)	Lactobacilli (cfu/g)
SD-1	3	3.8–3.9	10.5–11.5	1.8×10^5	5.4×10^4
SD-2	1–20	<4.0	~ 30	Min 10^6	Min 10^7

Source: Data obtained from supplier specifications.

2. Application and Properties

Commercial sourdoughs are ready to use, and are directly added (according to the recommended dosage) with the rest of ingredients and mixed together. Among the numerous functions, the addition of these sourdoughs confers flexibility and elasticity to the bread doughs. Their suppliers claim that they

Yield breads with a traditional taste and flavor
Produce crumbs with a natural color
Lengthen the shelf life of the bread

These sourdoughs have a different shelf life depending on their type, the liquid sourdough may be stored at 4 °C up to 3 months, while the powdered sourdough can be stored for 12 months at temperatures less than 25 °C.

3. Effects of Some Commercial Sourdoughs

Some of the commercial sourdoughs have been tested in the authors' laboratory in order to compare their effect and confirm their properties as sourdoughs. Their effect has been determined at the process and bread levels (Table 6).

Table 6 Some Characteristics of Bread Dough and Rolled Bread Formulated with Commercial Sourdoughs[a]

Sample	Stability (min)	Mixing tolerance index (BU)	Specific bread volume (mL/g)	Firmness (g)
—	4.8	70	4.1	720.7
SD-1	5.3	50	4.7	451.8
SD-2	4.8	60	4.6	430.7

[a] Stability and mixing tolerance indices were determined by the farinograph.
Source: Data from Rosell and Benedito (unpublished data).

Process Level. The addition of these sourdoughs resulted in:

Reduction of the dough development time in mixing stage. However, in opposition to the natural sourdough, use of these products did not decrease the dough stability. In one case high stability was obtained, along with a reduction of the mixing tolerance index.

Higher CO_2 production in addition to shortened proofing periods. The resulting bread doughs had lower pHs at the beginning and end of the fermentation time, as well as higher total titratable acidity than the bread doughs obtained from a straight dough process.

Bread Level. Breads containing these sourdoughs in their formulations were characterized by:

More intense flavor and better taste

Higher specific volumes, along with softer crumb and more uniform grain

In general, similar effects to natural sourdoughs were found in the rheological properties of doughs and the quality of the breads obtained when commercial sourdoughs were used.

C. Dehydrated Sourdoughs

The utility of these products is to confer taste and flavor without promoting active fermentation. They are the result of stabilizing and drying a mature sourdough under controlled conditions.

These are available in the market in two different forms:

A friable paste, similar to fresh baker's compressed yeast, that should be kept at refrigeration temperature (0–10 °C). Under these conditions, the product maintains its properties for 5 weeks after its manufacture.

Powdered products with a shelf life of at least 12 months in a sealed original package when stored under cool and dry conditions.

1. Preparation

Only one friable paste product is available in the Spanish market. The appearance is that of fresh baker's yeast. This product is prepared by fermentation in several stages. According to the producers, this sourdough is prepared by starting from an old sourdough, as a source of acidolactic bacteria (lactobacilli). Lactobacilli and yeast are prepared by separate fermentation, then combined in suitable proportions and added to flour and water for second-stage fermentation. At the end of the process, the fermented broth is stabilized, reinforced, and packaged.

The powdered products are prepared from a sourdough in a fermentation tank with flour and water as any spontaneous sourdough, after which they are submitted to a drying process. The time and temperature of the fermentation depend on the desired characteristics of the final sourdough. The drying process is a very important step that proceeds under strictly controlled conditions in a spray or a roller dryer depending on the acidity. A spray dryer for low-acidity products or a roller dryer for high-acidity products are applied.

The process is carried out in a biologically controlled way during several stages and until its complete maturation. During this process the most important compounds in the sourdough, such as short-chain organic acids, alcohols, CO_2, and other aromatic products, are produced. The product is a concentrated and stable sourdough, where the microorganisms are finally inactive due to the restricted medium conditions.

It is claimed by the producers that approximately 1 kg of this sourdough is equivalent to 10 kg of fresh sourdough. The recommended use level is 1–3% on the flour basis. In the market, there are different dehydrated sourdoughs classified according to both source (wheat or rye flour) and acidity (total titratable acids).

2. Effects of Dehydrated Sourdoughs

It is reported that the products:

1. Improve bread quality, produce good taste and flavor, improve crust color, increase bread volume, improve crust break, improve crumb texture, and extend the product shelf life.
2. Facilitate the breadmaking process, shorten the resting periods, improve dough machinability, strengthen the dough giving an equilibrium between tenacity and extensibility, increase water absorption, govern fermentation, enhance the tolerance of the dough to fermentation, and prevent the growth of undesirable microorganisms (bacteria, bacilli, and fungi).
3. Create better working conditions; shorten the time of work; assure uniform characteristics, easy and accurate dosage, and better control of the breadmaking process; and be immediately available for use without prior planning.

The product can be used in any type of breadmaking process and should be added at the beginning of the dough mixing stage, as indicated in Fig. 3.

The main difference between these products and fresh sourdoughs is that they are biologically inactive. The viability of the microorganism is lost during the stabilization process; in consequence, they may not be considered as a traditional good sourdough.

Some experiments carried out in the authors' laboratory with dehydrated sourdoughs of varied acidity reveal that the addition of this type of sourdough promotes changes in the rheological properties and baking performance of wheat breads.

DD Effect on the Rheological Properties. An increase of the water absorption (likely due to the presence of different alcohols and acids) (Table 7) was observed. The dough development time was not affected, in contrast to the diminishing effect produced with natural sourdough (Barber et al., 1981). In contrast, the use of these sourdoughs produces a decrease in the stability and an increase in the mixing tolerance index as was previously found with the natural sourdoughs (Barber et al., 1981).

A slight decrease in the resistance to deformation, extensibility, and deformation energy (measured by alveograph) of the dough was observed with the addition of dehydrated sourdough. These results are in agreement with previous findings of Barber et al. (1981) using natural sourdough.

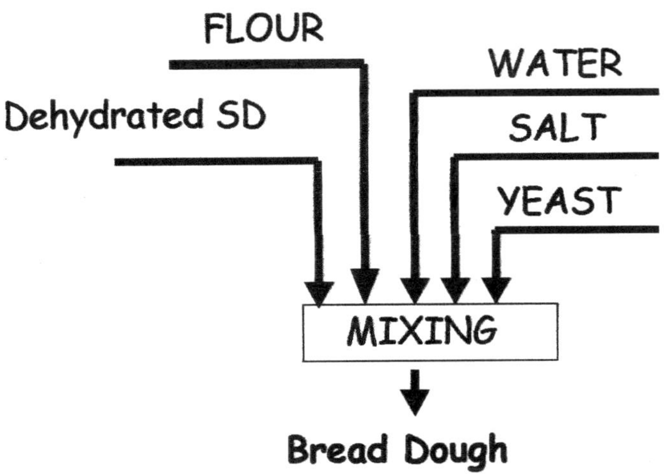

Figure 3 Use of dehydrated sourdough to obtain a final bread dough.

Table 7 Effect of Dehydrated Sourdough Addition on the Farinogram Characteristics of Wheat Doughs

Sample	TTA (mL NaOH 0.1 N)	Water adsorption (%)	Stability (min)	Mixing tolerance index (BU)
Without SD	—	56.7	6.2	38
SD-1	40	58.4	5.3	45
SD-2	70	57.9	4.7	55
SD-3	135	58.4	4.5	60

Source: Data from Rosell and Benedito (unpublished data).

Regarding effects on the fermentative properties (Table 8), all of the dehydrated sourdoughs tested produced an increase of the maximal dough volume but extended the proofing. The fermented dough in the presence of sourdoughs had lower stability. Dough stability decreased as the acidity increased, falling to zero at the highest acidity tested. CO_2 production increased with these sourdough additions with a slight decrease of the gas retention capacity. These results are also similar to the effect observed with the natural sourdoughs previously described by Barber et al. (1981, 1991a,b).

DD Effect on Baking Performance. (Table 9) Bread doughs formulated with these sourdoughs showed lower pH values both at the beginning and at the end of the fermentation, and the pH diminished with the sourdough acidity increase. This effect was similar to that observed by Barber et al. (1987, 1989a,b) and Martínez-Anaya et al. (1989) in bread dough formulated with sourdough from a multistage process. As the process progresses the acidity increases, and in consequence the final acidity of the bread dough also increases.

The amount of acetic and lactic acid increased with the acidity of the sourdough used in the formulation, but no differences were found in the amount before and after the fermentation, as was expected due to the absence of viable lactobacilli. The fermentation quotient (molar ratio lactic acid/acetic acid) ranged from 0 (when formulated without sourdough) to 2.5 when adding the sourdough of the highest acidity. In addition, the total titratable acidity in bread doughs was increased with the addition of these sourdoughs, and the extent of this effect was related to the sourdough's acidity.

The final breads, containing these sourdoughs, had higher specific volume and softer crumb than the bread formulated without sourdoughs.

Table 8 Analysis of Fermentation Stages of Doughs Containing Different Dehydrated Sourdoughs by Rheofermentometer[a]

Sample	TTA (mL NaOH 0.1 N)	Dough development: H_m (mm)	Dough development: Vol. loss (%)	Gas behavior: Total CO_2 (mL)	Gas behavior: Gas retention (%)
Without SD	—	44.9	42.3	2012	83.2
SD-1	40	52.2	58.3	2073	82.7
SD-2	70	51.5	60.3	2024	83.3
SD-3	135	48.4	100.0	2067	82.0

[a] H_m maximal dough height; Vol. loss drop in dough volume at the end of 3 h; gas retention volume of CO_2 retained in the dough at the end of the assay.
Source: Data from Rosell and Benedito (unpublished data).

From the above data it could be concluded that the overall effect of this type of sourdough is similar to that promoted by the use of a natural sourdough (Barber et al., 1981, 1985, 1989a,b, 1991a,b), with the advantage of being a ready-to-use product in a straight dough process. In addition, it is stable without requiring special storage conditions.

There are a number of dehydrated products marketed as substitutes for sourdoughs. According to the information given in suppliers' specification data, they are obtained by dehydration of fermented flour, poolish (liquid ferment used in France), flour dough, sourdough, or sponge dough, under controlled conditions. None of them are biologically active products.

Table 9 Some Dough and Bread Characteristics Affected by the Addition of Different Dehydrated Sourdoughs in the Formulation

Sample	Final pH in fermented dough	Volume (mL)	Specific volume (mL/g)	Firmness (g)
Without SD	5.2	429	5.0	899.7
SD-1	5.0	496	6.1	560.0
SD-2	5.0	542	6.7	561.4
SD-3	4.7	588	7.5	490.5

Source: Data from Rosell and Benedito (unpublished data).

All should be added at the beginning of the mixing time. Baker's yeast should be used, of course, as ingredient.

D. Substitutes for Sourdough

Most of the products of this type present on the market have been produced with the aim to reinforce the qualities of the flour, which is frequently very weak. They are composed by ascorbic acid, emulsifiers, and enzymes (similar to any breadmaking improver), in addition to different organic acids to impart acidity to the dough. There are a number of these products on the market commercialized by different producers and distributors.

In the authors' laboratory there have been carried out some experiments in order to compare the effects of various substitutes of sourdoughs available on the market (results unpublished). Changes produced on the rheological and fermentative characteristics of the dough as well as on bread quality are compared in Tables 10–12. Data have been obtained in the authors' laboratory during testing of some commercial substitutes of sourdoughs. Nevertheless, it has to be mentioned that there are large variations in the effect of these products due to their different compositions.

1. Effect on the Rheological Properties

The addition of these products yielded an increase of the water absorption, shortened the dough development time, and, in contrast to the natural and previous sourdoughs, caused an increase of the dough stability measured by the farinograph (Table 10).

In addition, a decrease of both dough resistance to deformation (P) and extensibility (L) measured by the alveograph was produced, and in consequence an increase of the ratio P/L. The energy required for the dough deformation (W) was also augmented. Therefore, the overall effect was an increase in the flour strength.

With respect to fermentation, Table 11 shows a large increase in the maximal dough volume, and this maximum is reached in a shorter time in the presence of these compounds. Different effects on the gas production were obtained with the substitutes tested. In some cases, the total CO_2 produced was higher than in the absence of the substitutes, whereas in others a reduction in gas production was observed. However, all of them promoted an increase of the gas retention coefficient.

Table 10 Effect on the Farinographic Characteristics of Wheat Doughs Added with Different Substitutes of Sourdough

Sourdough substitute	Water absorption (%)	Dough development time (min)	Stability (min)	Mixing tolerance index (BU)
—	54.6	2.5	8.5	40
1	55.5	2.5	10.0	30
2	55.6	2.0	10.0	30
3	55.5	2.0	13.5	10

Source: Data from Rosell and Benedito (unpublished data).

2. Effect on Breadmaking Process (Table 12)

The addition of sourdough substitutes in the bread formulation promoted a decrease of the pH during fermentation. Bread doughs reached higher volume than that without sourdough substitutes.

An increase of the bread volume was obtained with the addition of sourdough substitutes, although the extent of this increase was greatly dependent on the substitutes added. At the same time, there was a significant reduction of crumb firmness, obtaining breads with softer crumbs.

Experiments carried out adding some commercial substitutes of sourdoughs in the bread formulations reveal that dough rheology is modified in the same manner as the natural sourdough did. The effects on

Table 11 Effect on the Fermentation Stage Promoted by the Addition of Different Substitutes of Sourdough[a]

	Dough development		Gas behavior	
Sourdough substitute	H_m (mm)	T1 (min)	Total CO_2 (mL)	Gas retention (%)
—	38.5	102	1766	80.8
1	48.8	141	1688	83.6
2	70.5	141	1989	81.6
3	72.1	117	1891	83.4

[a] The measurements were determined by the rheofermentometer. H_m maximal dough height; T1 time required to reach maximal dough height; total CO_2 produced; gas retention volume of CO_2 retained in the dough at the end of the assay.
Source: Data from Rosell and Benedito (unpublished data).

Table 12 Some Bread Dough and Bread Characteristics Affected by the Addition of Different Substitutes of Sourdough in the Formulation

Sourdough substitute	Final pH in fermented dough	Volume (mL)	Firmness (g)
—	5.2	429	899.7
1	5.2	430	378.0
2	5.0	680	346.3
3	5.0	647	303.5

Source: Data from Rosell and Benedito (unpublished data).

both the baking performance and the final bread quality are similar to those observed with the use of natural sourdough. The extent of the effect was highly dependent on the nature and composition of the substitute of sourdough use in the formulation. However, the flavor and taste of the breads are not enhanced with these products.

VI. CONCLUSIONS

The purpose of adding sourdough to bread formulations is to enhance the bread quality attributes. Breads containing sourdough have higher volume, softer crumb, an intense flavor and taste, and extended shelf life. The substitutes of sourdough present in the Spanish market, despite their different nature and composition, led to bread doughs and breads with characteristics similar to those produced with the addition of natural sourdough. In addition to the natural benefits of the sourdoughs, these products are ready to use. They do not require any expertise and they ensure the uniformity of the product quality. The major difference among them is the intensity of their effect. The starters and the biologically active sourdoughs were the best products to reproduce the sourdough effects. However, the dehydrated sourdoughs have the advantage of extended shelf life and do not require special storage conditions; the unique difference with the previous products is that they retard the proofing. Finally, the last group of products, the substitutes of sourdough, are bread improvers but do not impart flavor and taste to the product.

It should be mentioned that commercial sourdoughs, as defined in this chapter, are coming into the Spanish market very slowly and are nowadays scarcely used by the Spanish bakeries. The products described are being applied mainly to special breads, where the enhancement of quality justifies an increase in prize.

The biologically active sourdough and the starters could find application in bread and bread product factories, counting with a qualified staff and quality control department. It is expected that the dehydrated sourdoughs will soon find an interesting application in small and medium-size bakeries as a substitute for the current sourdoughs formulated and used by bakers, which do not always have the desirable characteristics, effects, and reproducibility.

REFERENCES

Barber, B. (1992). Efectos de los prefermentos y de su combinación con técnicas de conservación, sobre los atributos sensoriales y características físicas y químicas del pan y sus cambios en el almacenamiento. Thesis dissertation, Universidad de Valencia.

Barber, B., Martinez, J., Cornejo, L. and Benedito, C. Effects of different sourdoughs on the rheological and fermentative characteristics of bread dough. Revista de Agroquimica y Tecnologia de Alimentos *31*: 512 (1991).

Barber, B., Ortolá, C. and Spicher, G. Lagerung von verpacktem Brot. II. Mitteilung: Einfluss der Zugabe von Saverteig und der Hitzebehandlung auf die Beschaffenheit des Weizenbrotes. Getreide Mehl Brot *44*: 235 (1990).

Barber, B., Ortolá, C., Barber, S. and Fernández, F. Storage of packaged white bread. III. Effects of sourdough and addition of acids on bread characteristics. Z. Lebensm. Unters. Forsch *191*: 442 (1992).

Barber, S. and Baguena, R. Microflora of the sourdough of wheat flour bread. V. Isolation, identification and evaluation of functional properties of sour dough's microorganism. Revista de Agroquimica y Tecnologia de Alimentos *28*: 67 (1988).

Barber, S. and Baguena, R. Microflora of the sourdough of wheat flour bread. XI. Changes during fermentation in the microflora of sourdoughs prepared by a multistage process and of bread doughs thereof. Revista de Agroquimica y Tecnologia de Alimentos *29*: 478 (1989).

Barber, S., Baguena, R. and Benedito de Barber, C. (1989). Microbiological and physicochemical characteristics of sourdoughs and their relation with the characteristics of bread doughs and breads thereof. In: Trends in Food Science. Proc. 7th World Congress of Food Sci. Technol. (A. H. Ghee, L. W. Sze, F. C. Woo, eds.). Singapore Inst Food Sci Technol, pp. 187–191.

Barber, S., Baguena, R., Benedito Barber, C. and Martínez-Anaya, M. A. Evolution of biochemical and rheological characteristics and breadmaking quality during a multistage wheat sourdough process. Z. Lebensm. Unters. Forsch *192*: 46 (1991).

Barber, S., Baguena, R., Martínez-Anaya, M. A. and Torner, M. J. Microflora of the sourdough of wheat bread. I. Identification and functional properties of microorganisms of industrial sourdoughs. Revista de Agroquimica y Tecnologia de Alimentos *23*: 552 (1983).

Barber, S., Benedito de Barber, C. and Planells, V. The sponge for breadmaking. Its influence on the rheological and fermentative characteristics of wheat flour doughs. Revista de Agroquimica y Tecnologia de Alimentos *20*: 403 (1981).

Barber, S., Benedito de Barber, C., Martínez-Anaya, M. A. Martínez, J. and Alberola, J. Changes in volatile organic acids C_2–C_5 during fermentation in bread doughs containing commercial starters and pure cultures of microorganisms. Revista de Agroquimica y Tecnologia de Alimentos *25*: 223 (1985).

Barber, S., Martínez-Anaya, M. A. and Baguena, R. Microflora of the sour dough of wheat flour bread. II. Functional properties of commercial yeast and pure strains of S. Cerevisiae thereof in sugar solutions. Revista de Agroquimica y Tecnologia de Alimentos *25*: 436 (1985).

Barber, S., Martínez-Anaya, M. A. and Baguena, R. Microflora of the sour dough of wheat flour bread. III: Functional properties of commercial yeasts and pure strains of S. Cerevisiae thereof, in wheat flour dough. Revista de Agroquimica y Tecnologia de Alimentos *25*: 447 (1985).

Barber, S., Martínez-Anaya, M. A., Baguena, R. and Torner, M. J. Microflora of the sour dough of wheat flour bread. IV. Functional properties of microorganism isolated from commercial wheat flour sourdoughs. Revista de Agroquimica y Tecnologia de Alimentos *27*: 107 (1987).

Barber, S., Torner, M. J., Martínez-Anaya, M. A. and Benedito, C. Microflora of the sourdough of wheat flour bread. IX. Biochemical characteristics and baking performance of wheat doughs elaborated with mixtures of pure microorganisms. Z. Lebensm. Unters. Forsch *189*: 6 (1989).

Collar-Esteve, C., Benedito de Barber, C. and Martínez-Anaya, M. A. Microbial sourdoughs influence acidification properties and breadmaking potential of wheat dough. J Food Sci. *59*: 629 (1994).

Collar, C., Andreu, P. and Martínez-Anaya, M. A. Interactive effects of flour, starter and enzyme on bread dough machinability. Z. Lebensm. Unters. Forsch *207*: 133 (1998).

Collar, C., Mascaros, A. F., Brito, M. and Benedito de Barber, C. Contribution of the microbial mass to the nitrogen profile of wheat bread doughs started with pure and mixed cultures of yeast and lactic acid bacteria. Z. Lebensm. Unters. Forsch *193*: 332 (1991).

Collar, C., Mascaros, A. F., Frechilla, S. and Benedito de Barber, C. Protein capability of starting-doughs lactic acid bacteria. Revista de Agroquimica y Tecnologia de Alimentos. *31*: 343 (1991).

Collar, C., Mascaros, A. F., Prieto, J. A. and Benedito de Barber, C. Changes in free amino acids during fermentation of wheat doughs started with pure culture of lactic acid bacteria. Cereal Chem. *68*: 66 (1991).

Galal, A. M., Johnson, J. A. and Varriano-Marston, E. Lactic and volatile (C_2–C_5) organic acids on San Francisco sourdough french bread. Cereal Chem. *55*: 461 (1978).

Gobbetti, M., Simonetti, M. S., Corsetti, A., Santinelli, F., Rossi, J. and Damiani, P. Volatile compound and organic acid productioins by mixed wheat sourdough starters: influence of fermentation parameters and dynamics during baking. Food Microbiol. *12*: 497 (1995).

Hansen, A. and Hansen, B. Flavour of sourdough wheat bread crumb. Z. Lebensm. Unters. Forsch *202*: 249 (1996).

Martínez-Anaya, M. A. Carbohydrate and nitrogen related components in wheat sourdough processes: a review. Adv. Food Sci. *18*: 185 (1996).

Martínez-Anaya, M. A. Review: Factors influencing the quality of wheat sourdough processes and the use of chemometrics in quality assessment. Revista Española de Ciencia y Tecnologia de Alimentos *34*: 469 (1994).

Martínez-Anaya, M. A., Benedito de Barber, C. and Collar-Esteve, C. Effect of processing conditions on acidification properties of wheat sourdoughs. Int. J. Food Microbiol. *22*: 249 (1994).

Martínez-Anaya, M. A., Pitarch, B. and Benedito de Barber, C. Microflora of the sour dough of wheat flour bread. XIII. Multivariate data analysis applied to the selection and classification of microbial starters for bread dough fermentation. Revista de Agroquimica y Tecnologia de Alimentos *31*: 366 (1991).

Martínez-Anaya, M. A., Pitarch, B., Bayarri, P. and Benedito de Barber, C. Microflora of the sourdough of wheat flour bread. VIII. Changes in sugars and organic acid production during fermentation of wheat doughs made with pure strains of microorganisms. Revista de Agroquimica y Tecnologia de Alimentos *29*: 63 (1989).

Martínez-Anaya, M. A. (1983). Modificaciones de las propiedades funcionales de la masa madre y de la masa panaria durante la fermentación y efectos de los microorganismos implicados. Thesis dissertation, Universidad de Valencia.

Martínez-Anaya, M. A., Devesa, A., Andreu, P., Escriva, C. and Collar, C. Effects of the combination of starters and enzymes in regulating bread quality and shelf life. Food Sci. Technol. Int. *4*: 425 (1998).

Martínez-Anaya, M. A., Llin, M. L., Macias, M. P. and Collar, C. Regulation of acetic acid production by homo- and heterofermentative lactobacilli in whole-wheat sourdoughs. Z. Lebensm. Unters. Forsch *199*: 186 (1994).

Martínez-Anaya, M. A., Torner, M. J. and Benedito, C. Microflora of the sour dough of wheat flour bread. XIV. Changes in volatile compounds during fermentation of doughs prepared with pure microorganisms and their mixtures. Z. Lebensm. Unters. Forsch *190*: 126 (1990).

Mascaros, A. F., Martinez, C. S. and Collar, C. Metabolism of yeasts and lactic acid bacteria during dough fermentation relating functional characteristics of fermented doughs. Revista Española de Ciencia y Tecnologia de Alimentos *34*: 623 (1994).

Oura, E., Suomalainen, H. and Viskari, R. Breadmaking. In: Fermented Foods, Academic Press, New York, 1982, p. 87.

Rouzaud, O. and Martínez-Anaya, M. A. Effect of processing conditions on oligosaccharide profile of wheat sourdoughs. Z. Lebensm. Unters. Forsch *197*: 434 (1993).

Rouzaud, O. and Martínez-Anaya, M. A. Relationships between biochemical and quality-related characteristics of breads, resulting from the interaction of flour, microbial starter and the type of process. Z. Lebensm. Unters. Forsch *204*: 321 (1997).

Spicher, G. and Stephan, H. (1982). Handbuch Sauerteig, Biologie, Biochemie, Technologie. BBV. Wirtschafts Informationen GmbH, Hamburg.

Spicher, G. Baked Goods. (1982). In: Biotechnology, Vol. 5 (Rehm, H. J. and Reed, G. eds.). Verlag Chemie, Weinheim.

Tanaka, K., Furukawa, K. and Matsumoto, H. The effect of acid and salt on the farinogram and extensigram of dough. Cereal Chem. *44*: 675 (1967).

Torner, M. J., Bainotti, A., Martínez-Anaya, M. A. and Benedito de Barber, C. The microflora of the sour dough of wheat flour bread. XII. Effect of freezing on the viability and functional properties in wheat flour doughs of microbial mass from lactic acid bacteria. Z. Lebensm. Unters. Forsch *189*: 554 (1989).

Torner, M. J., Martínez-Anaya, M. A., Antuna, B. and Benedito de Barber, C. Headspace flavour compounds produced by yeasts and lactobacilli during fermentation of preferments and bread doughs. Int. J. Food Microbiol. *15*: 145 (1992).

Tsen, C. C. A note on effects of sulphydryl groups and rheological properties of dough and its implication with the sulphydryl-disulfide interchange. Cereal Chem. *43*: 456 (1966).

Wehrle, K. and Arendt, E. K. Rheological changes in wheat sourdough during controlled and spontaneous fermentation. Cereal Chem. *75*: 882 (1998).

10

Preferments and Sourdoughs for German Breads

Klaus Lorenz
Colorado State University, Fort Collins, Colorado, U.S.A.

J.-M. Bruemmer
Federal Center for Cereal, Potato, and Lipid Research in Detmold & Muenster, Detmold, Germany

I. INTRODUCTION

Preparations of preferments and sourdoughs by various methods available to the German baker pursue the same goal but are given different degrees of emphasis to give bakery products certain quality characteristics. This has led to the development of widely differing processing techniques to cover a wide variety of products taking into consideration the availability and the types of raw material, in particular the types of flours [1].

For breads made predominantly with rye flour, sours are expected to produce a good crumb moisture, pore size, and elasticity; impart an aromatic and pleasing flavor; improve overall eating quality and shelf life.

For breads made predominantly with wheat flour, a sour is mainly expected to improve flavor and shelf life. Sourdoughs in baked products made from low-extraction (low-ash) wheat flours do not produce the same beneficial effects as they do in rye-based products.

All doughs containing rye flour have to be acidified. Sour conditions have a positive influence on the swelling power of the pentosans and mucilages of rye flour and at the same time inactivate some enzymes, particularly amylase. The degree of acidification depends on the proportion of rye flour in the flour blend [2].

There is a general trend toward simplifying and automating fermentation processes making possible the use of computer-aided process control systems and a "computer-integrated manufacture." Economic pressures force bakeries to be more productive and at the same time retain or improve the quality of their products by using the latest technology.

Automation of fermentations presents a difficult problem since it involves a complex mass flow that depends to a large extent on biological laws. Specific variables that affect microbiological reaction processes during fermentation times, which can vary from 1 to 24 h, have to be carefully controlled because they affect final product quality [1].

Sourdough fermentation involves a considerable number of heterogeneous metabolic and fermentation reactions, which can not be discussed in this chapter.

Microbiological sourdough research started in Germany around 1900. Seibel and Bruemmer [2] cite the various researchers and their publications over the early years to the present. The research data are contained in a handbook published by Spicher and Stephan [3]—the only text book devoted to sourdough written in German.

II. PREFERMENTS

Preferments in Germany are generally considered for 100% wheat-based bakery products using yeast as starter. They are fermented for a specific time period using a portion of the total flour before preparation of the dough. Fermentation times can be as short as 4 h or as long as 14–16 h overnight. Even the long preferments hardly intensify or improve product flavor. However, they affect, crumb softness, structure, and pore size. Whenever the yeast starter suppresses the spontaneous flora of the raw material flour, there is no acidification.

Table 1 Preferments for Wheat-Based Bakery Products

Fermentation time (h)	4	16
Temp. of preferment (°C)	26	22
Yeast (% of preferment flour)	2	1

Typical preferment parameters are given in Table 1. The absorption will depend on the working conditions in the bakery, i.e., is the preferment going to be handled by hand after fermentation or is it going to be pumped? Addition of salt or gluten is unnecessary and may even reduce the advantages of a preferment.

Baked goods with high fat (15–45%) and sugar (10–20%) contents can also benefit from preferments. Crumb structure and characteristics are improved. Preferment fermentation times vary from 1 to 4 h at a temperature of 25 °C using 30% of the total flour. The entire amount of yeast called for in the formulation is used in the preferment. Absorption also depends on working conditions in the bakery as stated above.

Experimental baking trials have shown that formulation critically determines dough characteristics, machineability, color development and crumb structure, flavor, taste and eating quality of products made with high fat and sugar levels. Preferments favorably affect final proofing, product volume, and surface characteristics as well as crumb softness. Flavor improvements generally are not noticeable [4,5].

III. RYE SOURDOUGHS

Sourdoughs are either liquid, paste-like, or dough-like depending on the ratio of flour to water. The consistency of sourdough decreases during the time of ripening. The ratio of flour to water is chosen depending on whether the dough is transferred by hand or by pump from fermenter to mixer.

Different sourdough processes are used for the production of bread containing rye flour. Some are based on experience and are of purely empirical origin, whereas others are based on microbiological studies. Three-, two-, and one-stage sourdough processes are used [2,6–8].

A. Multiple-Stage Sourdough

In this procedure, a "starter," which is taken from a ripe sour, is activated by the addition of rye flour and water. The temperature of this "Anfrischsauer" should be 22–24 °C, which is favorable for sourdough yeasts. The ripening time is 3–6 h. At the end of the ripening time, the Anfrischsauer is increased to a basic sour by again adding rye flour and water. The temperature of the basic sour should be 25–28 °C and the ripening time 15–24 h. A full sour is finally prepared at a temperature of 30–33 °C and is permitted to ripen for 3 h. A starter is held back from the full sour for the following day's production. No yeast is used in the multiple stage method [8].

A typical multiple-stage overnight sourdough procedure is outlined in Table 2. This method is undoubtedly the classical one. It is rather complicated and probably too cumbersome for bakeries that have had little experience in rye bread production.

B. Two-Stage Sourdough

In the two-stage procedure, the "starter" is used directly to prepare a basic sour. The conditions of the basic sour are similar to the second stage of the multiple-stage procedure. Preparation of the full sour is as described above. Some baker's yeast is added to the final dough to provide some additional leavening.

A typical two-stage sourdough procedure is outlined in Table 3.

C. One-Stage Sourdough

One-stage sourdough procedures have been developed to simplify the classical multiple-stage method. There are three different one-stage sourdough methods that a baker may use:

1. Detmold one-stage sour [9]
2. Monheim salt sourdough [10]
3. Berlin short sourdough [11]

Ripening times of these one-stage sourdoughs range from 3.5 to 24 h with dough temperatures ranging from 24 °C to 35 °C [9–12].

The Detmold one-stage sour requires a relatively long ripening time of 15–20 h. Sourdough temperatures should be 25–27 °C and the dough yield (combined amounts of water and flour) about 180. The Monheim salt method requires a high temperature (30–35 °C) for a long time (18–24 h) since the addition of salt to the sour affects the metabolism of the organisms. The dough yield is about 200. Higher amounts of yeast are required in the final dough because of a lower pH of this sourdough. The process is easy to control and produces a higher yield of dough and bread than other sourdough processes due to smaller fermentation losses. The Berlin sourdough method is the only short–time method. Sourdough temperatures are about 35 °C with ripening times of 3–4 h and a dough yield of 190.

With each one-stage method, a small amount of yeast is added at the dough stage. The procedures have been automated and the sourdoughs can now be stored at reduced temperatures for a certain time without decreasing their acid formation properties.

Table 2 Multiple-Stage Sour Overnight Procedure[a]

	Starter (kg)	Rye flour (kg)	Water (L)	Sour or dough yield[b]	Temperature initial	(°C) Final	Ripening time (h)	Time schedule
Anfrischsour	0.5	1.5	2.0	230	22	23	6	8 am
Basic sour	—	3.0	2.0	190	25	27	6	2 pm
Full sour[c]	—	40.5	32.0	180	23	25	6	8 pm
Dough[d]	—	55.0	39.0	175	30	32	—	4 am

[a] Total flour = 100 kg; 100% rye flour; 45% of rye flour included in sour.
[b] Dough yield = percent based on flour weight.
[c] Starter for Anfrischsour is taken from full sour at end of ripening time.
[d] Salt addition at dough stage = 1.6% based on total flour.
Source: Adapted from Ref. [8].

Table 3 Two-Stage Sourdough Procedure[a]

	Starter (kg)	Rye flour (kg)	Wheat flour (kg)	Water (L)	Sour or dough yield[b]	Temp. (°C)	Ripening time (h)
Basic sour	1.0	16.0	—	8.0	150	23–27	15–24
Full sour[c]	—	24.0	—	24.0	180	33–29	3
Dough[d]	—	50.0	10.0	38.0	170	29	—

[a] Total flour = 100 kg; 90% rye + 10% wheat flours; 40% of total flour included in sour.
[b] Dough yield = percent based on flour weight.
[c] Starter taken from full sour at end of ripening time.
[d] Yeast (1.8%) and salt (1.8%) added at dough stage.
Source: Adapted from Ref. [8].

Typical one-stage sourdough procedures for mixed rye–wheat breads are outlined in Table 4. The wheat flour is always added at the dough stage. A typical Berlin short sour overnight procedure for the production of 100% rye flour breads is given in Table 5. This method has become very popular in Europe. It saves time, is short and simple, and the sour is relatively easy to maintain, thus assuring uniform bread quality from day to day.

The basic differences between the multiple-stage procedure and the Berlin sourdough method are summarized in Table 6.

D. Rules for Dough Acidification

General rules for dough acidification are as follows [13]:

High % of rye in formulation = greater acidification needed
Low % of rye in formulation = less acidification needed
Soft doughs = greater acidification needed
Firm doughs = less acidification needed
Big loaves (>1500 g) = greater acidification needed
Small loaves (500–750 g) = less acidification needed
Flour with high enzyme activity = greater acidification needed
Flour with low enzyme activity = less acidification needed

E. Typical Acid Values of Breads

Typical acid values of breads are as follows [2]:

Rye breads (at least 90% rye) = 8.0–10.0
Mixed rye breads (51–90% rye) = 7.0–9.0
Mixed wheat breads (<50% rye) = 5.0–8.0

Acid value is determined by titrating 10 g of suspended bread with 0.1 N NaOH. For acid values in sourdoughs, "Teigausbeute" (TA) or dough yield (combined amount of water and flour, flour being 100) has to be considered. The amount to be titrated is one twentieth of the TA. If TA of the sourdough is 200 (100 parts of flour and 100 parts of water), 10 g (200 divided by 20) is used for titration; if TA is 250 (100 parts of flour and 150 parts of water), 12.5 g (250 divided by 20) is used for titration with 0.1 N NaOH to a pH value of 8.5. In each case 5 g of acidified flour is used in the titration [2].

F. Current Trends

There is a trend toward one-stage sourdoughs. The acidified amount of flour in rye (90% rye) and rye-mixed breads (50–89% rye) ranges from 30% to

Table 4 One-Stage Sourdough Procedures[a]

Sourdough process	Starter[b] (kg)	Rye flour (kg)	Wheat flour (kg)	Water (L)	Yeast (kg)	Salt (kg)	Sour or dough yield[c]	Sour or dough temp. (°C)	Sour ripening time (h)
Detmold one-stage sour	1.5	30.0	—	24.0	—	—	180	25–27	15–20
	—	35.0	35.0	45.0	1.9	1.9	170	26	—
Monheim salt sour	5.0	25.0	—	25.0	—	0.5	200	33–23	15–30
	—	40.0	35.0	42.0	2.0	1.5	154	26	—
Berlin short sour	8.0	40.0	—	36.0	—	—	190	35	3.4
	—	25.0	35.0	32.0	1.8	1.8	168	28	—

[a] Total flour = 100 kg; 65% rye and 35% wheat flours; all wheat flour added at dough stage.
[b] Starter taken from sour at end of ripening period.
[c] Dough yield = % based on flour weight.
Source: Adapted from Ref. [8].

Table 5 Berlin Short Sour Overnight Procedure[a]

	Starter[b] (kg)	Flour (kg)	Water (L)	Sour or dough yield[c]	Temp. initial	(°C) Final	Time schedule
Sour	17.0	50.0	35.0	170	30	35	8 pm
Dough[d]	—	50.0	30.0	165	—	32	5 am

[a] Total flours = 100 kg; 100% rye flour.
[b] Starter taken from sour at end of ripening period.
[c] Dough yield = % based on flour weight.
[d] Salt (1.5 kg) and yeast (0.5–1.0 kg) added at dough stage.
Source: Adapted from Ref. [8].

Table 6 Basic Differences Between Multiple-Stage and Berlin Sourdough Method

	Multiple-Stage			Berlin Short Stage	
Purpose	Yeast propogation for aeration of dough Acid formation to improve flour baking quality Bread flavor development Hydration of portion of flour			Acid formation Flour hydration	
Number of stages	3			1	
	Stage	**Basic Sour Overnight**	**Full Sour Overnight**	**Day Procedure**	**Overnight Procedure**
Dough or sour consistency	Anfrischsour	= soft	soft	soft	medium
	Basic sour	= stiff	medium	to	to
	Full sour	= medium	stiff	medium	stiff
	Stage	**Basic Sour Overnight**	**Full Sour Overnight**	**Day Procedure**	**Overnight Procedure**
Temp. (°C) of stages	Anfrischsour	= 24°–26°	24°–26°		
	Basic sour	= 22°–26°	26°–28°	34°–36°	28°–32°
	Full sour	= 28°–32°	24°–28°		
	Different for each stage			**Day Procedure**	**Night Procedure**
Time of sour ripening	Total time = 18–24 h			2–3 h	6–9 h
Characteristic and condition of sour	Volume enlarged, shape rounded			Little change in volume and shape, appears lifeless	
Yeast addition	Not absolutely necessary—If added, 0.5–1.0%			1–2% necessary	
Fermentation losses	2–2.5% based on flour			1–1.2% based on flour	

Source: Adapted from Ref. [8].

35%, while in wheat breads (90% wheat) the range is from 5% to 10%. In wheat whole-grain breads 10% of the flour is acidified whereas in mixed wheat breads (<50% rye) 20% of acidified flour as sour is added to the dough.

To obtain the best possible crumb characteristics (consistency, pore size, elasticity), 1.5–1.8% yeast is used in rye-dominant breads, 3% in wheat and whole wheat breads, and about 5% in small wheat-baked products.

It has become customary not to prepare a separate sour for 100% wheat-based breads products, but to use about 3% of a rye sourdough. The resulting slightly darker breads crumb is appreciated rather than rejected in Germany [13].

G. Factors Affecting Sourdoughs

Sourdough fermentations are affected by (1) flour type, (2) starter propagation, (3) absorption and dough temperature, (4) atmospheric oxygen, and (5) pH.

1. Flour Type

The type of flour used in a sourdough greatly influences the ripening time of the sour. More acid is required to reach the critical pH value for optimal rye bread quality when a high-extraction flour is used than when a low-extraction flour is used. Due to differences in buffering capacity, the growth of microorganisms and formation of acids proceeds much more slowly in a dough prepared with whole rye meal than in a dough made from a low-extraction flour [8].

2. Starter Propagation

The starter of a sourdough is normally taken from the ripe sour and can be stored in its original form for several days without biological deterioration.

Pure culture sours can be purchased initially or the baker can produce a starter by spontaneous fermentation relying on airborne microorganisms and the bacteria indigenous to the flour. By preparing a dough from rye flour and water using a ratio of 1:1 and refreshening this dough daily for several days, a normal starter can be produced.

Proper propagation of the starter is very important. Rapid bacterial growth is obtained in a dough which is only slightly acidic initially by mixing the starter with a large amount of flour and water (sour one-stage procedure). This assures continuous formation of desired acids. A smaller dough volume increase at the next stage of the sourdough process (dough stage) causes the critical pH value to be reached in a short time due to the greater reservoir of bacteria and acids. It brings to a standstill the initially rapid rate of acidification and favors the growth of yeasts [8].

3. Absorption and Dough Temperature

High absorption (soft doughs with a "Teigausbeute" (TA) of 200–300) leads to faster bacterial growth and stronger fermentation, but fermentation tolerance is low. Low absorption (firm doughs with a Teigausbeute of 150) will retard bacterial activity. These doughs have a high fermentation tolerance [13]. Absorption levels over 100% (TA 200) will cause a decrease in acidification, however, due to reductions in the concentrations of nutrients as the result of the increase in dough dilution [8].

Dough consistency also affects the ratio of lactic to acetic acid in a sour. Soft doughs (TA 200–300) favor lactic acid formation, whereas stiff doughs (TA 150) favor the formation of acetic acid.

The critical pH value for bacterial activity, i.e., the pH value at which bacterial cell multiplication stops, is 4.0. When this level is reached, there is a great decline in the intensity of acidification. This pH value is reached in 3 h at 100 °F, in 4.5 h at 86 °F, and in about 8 h at 77 °F. The most favorable conditions for acidification of a sour are a temperature of 95 °F and an absorption of 90% (TA 190) [8].

4. Atmospheric Oxygen

Atmospheric oxygen has a definite effect on the most important bacteria in a sourdough and on acid formation. It retards the growth of *Lactobacillus plantarum* but slightly promotes the growth of *Lactobacillus brevis* [3].

5. pH

The pH reached by the sourdough during fermentation is of utmost importance. It influences the hydration capabilities of the rye flour, the activity of the enzymes in the dough system, and consequently the pH, grain, texture, crust color, flavor, and shelf life of the bread.

Careful maintenance of the starter and proper temperature and absorption of the sour for an appropriate length of time with a given rye flour will assure that the correct pH is reached during fermentation for optimal bread quality. Desirable bread pH values are as follows: for bread baked with 100% rye flour, 4.2–4.3; for bread baked with 70% rye flour and 30% wheat flour, 4.4; and for bread baked with 30% rye flour and 70% wheat flour, 4.65–4.75. Higher pH values produce unsatisfactory crumb elasticity [8].

IV. WHEAT SOURDOUGHS

Wheat sourdoughs have been thoroughly studied at the German Federal Research Institute for Cereal and Potato Processing. Investigations have included studies of the effects of different water absorptions (60, 100, and 150% based on wheat flour, TA 160, 200, 250), dough temperature (25, 30, and 35 °C), percentages of "Anstellgut"—a nontranslatable German word referring to the starter culture from a ripe sourdough (1, 5, 10, and 20%), length of sourdough ripening time (up to 24 h) with wheat flour of 0.55% and 1.05% ash, respectively, as well as whole wheat flour (ash 1.5%) [14–18].

A. Effect of Flour Type

In wheat sourdoughs, the final acid value after 24 h of ripening increased from 13 to 23 with increasing ash content of the flour from 0.55 to 1.5. Sourdough pH values show less variation and fewer differences than acid values. However, pH values below 4.0 are reached more rapidly with wheat flours of low ash content. The warmer and softer the dough, the lower the pH.

B. Effect of Anstellgut on Sour Acidification

Use of 0.5–5.0% of Anstellgut in the sour produced slower initial acid formation. Higher levels produced lower final acid values. Anstellgut amounts of 1–5% in the sour are generally recommended for the development of wheat sourdoughs. This does not cause an initial rapid pH drop; it allows sufficient development of active microorganisms in the sourdough and prevents possible unwanted fermentation.

C. Effect of Temperature on Final Acid Value

The effect of temperature on final acid value is similar in all sourdoughs. Final acid value increased as sourdough temperature increased from 25 °C to 30 °C. In contrast to sourdoughs prepared with flour of 0.55% ash, flours higher in ash reached an even higher final acid value as sourdough temperature increased to 35 °C. Highest acid values were obtained with warm (30–35 °C) and soft (absorption 150–200%) sourdoughs as well as with intermediate amounts of Anstellgut (1–5%). The lowest acid values were obtained with an absorption of 60%. A sourdough temperature of 25 °C also produced low final acid values. The amount of Anstellgut used in the sour had no effect on total acidification with ripening times of up to 24 h.

However, short ripening times affect acid levels considerably, because of differences in speed of acidification [5,14–17].

D. Effect of Absorption on Final Acid Value

Sourdough absorption is a very important factor in the production of wheat sourdoughs. Independent of the type of flour used, an increase in final acid value was measured with an increase in absorption. The difference in acid value between the firmest sourdough (absorption 60%) and the softest (absorption 150%) was 3–5 acid degrees when using wheat flours, whereas whole wheat flour with absorptions of 150% or 200% did not produce additional increases in acid value.

E. Effect of Sourdoughs on Bread Quality

Wheat sourdough has characteristics similar to those of rye sour if the flour used has approximately the same ash content.

If ripening of the sourdough is scheduled overnight, use of a wheat flour with up to 1% ash is recommended, as are high absorption, high dough temperature, and an intermediate amount of Anstellgut. Use of 7.5–10% of the wheat flour in the sourdough will produce a pleasantly aromatic wheat bread aroma and flavor.

Table 7 shows quality characteristics of wheat sourdough bread made from different flours.

V. COMMERCIAL SOURDOUGH STARTERS

New sourdough starters have been developed in the last few years based on many years of experience [18–21]. The initial idea was to avoid using in a bakery a "spontaneous" sour or "Anstellgut" with an uncertain microbiological composition or even containing undesirable microorganisms.

Studies of the sourdough flora and the influence of different kinds of lactic acid bacteria on bread quality [3] have led to the development of various microbiological starters with defined numbers and species of microorganisms. These starters contain only a few species of lactic acid bacteria and no yeasts. As a consequence, time, temperature, dough yield, and starter quantity have to be controlled more stringently than with more varied starters. Fermentation may go wrong if conditions optimal for a particular lactic acid bacterium are not maintained. There is less danger when multiform starters are used because if some bacterial species do not

Table 7 Effect of Sourdough on Quality of Wheat Sourdough Bread[a]

Flour ash (%)	0.45	0.55	0.81	0.99	1.05
Absorption (%)	57	57	57	57	57
Dough characteristic					
Top	Normal	Slightly wet	Slightly wet	Slightly wet	Slightly wet
Elasticity	Normal	Slightly elastic	Slightly elastic	Slightly elastic	Slightly elastic
Volume (mL/100 g flour)	610	600	600	580	590
Crumb structure	Tender	Slightly tender	Slightly tender	Somewhat coarse	Somewhat coarse
Crumb elasticity	Good	Good	Good	Good to somewhat tough	Good to somewhat tough
Taste	Mild	Mild, aromatic	Aromatic	Aromatic, sour	Aromatic, sour
Acid value/pH	2.3/5.6	3.1/5.0	3.7/4.8	4.2/4.7	3.9/4.8
Crumb color	Normal	Normal	Normal	Slightly darker	Slightly darker

[a] 10% of sourdough used to bake breads.
Source: Adapted from Refs. [15–18].

encounter optimum conditions, others will find the environment sufficiently favorable for proliferation.

Liquid or semiliquid and dried or lyophilized starters have been developed, intended to be used to acidify a bread or roll dough directly. Such starters are not used for new sourdoughs. For these sourdoughs, microorganisms were selected not only on the basis of their acidifying power but also on their drying resistance, which affects survival rate. Because of drying, reactivation of the lactic acid bacteria takes longer than reactivation of active dry baker's yeast. Dried sourdoughs presently have only low acid-forming capacities in bread and roll doughs. Breads and rolls with a high proportion of rye flour still tend to taste different when produced with a dry sour compared with baked goods manufactured with a freshly prepared sourdough in the bakery. For acidifying wheat flour doughs, the intrinsic acidity and reactivation capacity of the products is sufficient.

The pure culture or mixed-culture starters available today have some definite advantages but also some disadvantages. These cultures remain microbiologically stable in a relatively wide temperature range with a high level of activity and adapt well to changes in processing parameters. However, liquid or semiliquid cultures remain active and stable for only 4 months under cool storage. Stable for a longer time are dried or lyophilized pure or mixed cultures. However, these require a relatively long reactivation time, making direct and immediate use as sourdough starters unfeasible. Table 8 provides a list of such starters, their manufacturers, and then applications.

VI. SOURDOUGH EQUIPMENT

There have been attempts to simplify fermentation and maturing of sourdough and to increase the safety of the process by using machines since the end of the 1970s in Germany.

Sourdough production machines became important due to the increased consumption of sourdough bread. German bakeries then began to replace the multistage sourdough processes by single-stage ones that are more suitable for mechanized sourdough production. Modern bakeries now attach as much importance to sourdough production equipment as to investments in the modernization and mechanization of flour storage, dough making, fermentation, and baking. Increased efforts are being made to develop new types of sourdough equipment, improve automatic control systems, and supply more information about the proper use of this equipment.

Table 8 Commercially Available Starters

Manufacturer	Product name	Application
Ireks GmbH—Kulmbach	"Vollsauer"	Rye breads
	"Weizensauer"	Wheat breads
Ulmer Spatz—Ulm	"Ulmer Anfrischsauer"	Rye, mixed breads
Diamalt	"Pasten Malzsauer"	Rye breads
Gebr. Jung GmbH—Frankfurt	"Sauerteigpflege"	Rye, mixed breads
Boehringer—Ingelheim	"Vollsauer"	Rye, mixed breads
	"Sauer Controller"	Rye breads
	"Flintbeker Weizensauer"	Wheat breads
Uniferm—Werne	"Fermentsauer, Roggen"	Rye, mixed breads
	"Fermentsauer, Weizen"	Wheat breads
	"Vollkorn Sauerteig"	Mixed breads
Isernhaeger Landkost Menge	"Profermenta"	Rye, mixed breads
Boecker—Minden	"TK Starter"	San Francisco sour breads
	"Silo Starter"	Rye, mixed breads
	"Bio Starter"	Rye breads

Source: Ref. [22].

A. Description of Equipment

Machines for the production of sourdough now take care of operations such as dough processing, sourdough fermentation, and metering up to automatic discharge. All consist of a single-wall vessel with a lid and a stirrer, the speed of which can be varied. Depending on the volume required, the ratio between the height and diameter will range from 0.8 to 1.3, with smaller vessels having relatively larger diameters. They are emptied by direct evacuation or with a hand pump, and all are suitable for production of rye and wheat sourdoughs [20,22]. The transfer of mature sourdoughs for the preparation of bread and roll doughs is possible since the machines are only used for sourdoughs that can be pumped. Dough yields are of the order of 225 or more (at least 125 L of water is added per 100 kg of a rye milling product). Usually such soft sourdoughs ferment more rapidly than firmer ones and have a higher lactic acid content [18].

After normal ripening, these soft sourdoughs have a low standing tolerance; their acidity continues to increase rather rapidly. To prevent this, some machines have been equipped with cooling equipment.

After completion of the ripening process, some machines automatically retain a certain amount of fermented sourdough as starter for the next sourdough batch [20].

Different sourdough machines for medium and large-scale production have been tested. There is general agreement that a production line should have the following:

- Automatic weighing and metering of water, milling products, and other ingredients
- Computer-controlled presetting of time, temperature, dough yield
- Stirrers with adjustable speed for continuous or cyclical operation (crusher for coarse ingredients, if required)
- Temperature control, in particular intensive cooling for vessel wall and floor elements on the basis of adjustable time or pH/acidity value (very important if raw materials with different buffering capacity are used)
- Equipment for total evacuation of the system for cleaning or for partial evacuation with the starter remaining in place
- Metering of constant and variable starter quantities (controlled by pH/acidity value)
- Easy cleaning and sterilization (steam lines, if required)
- Automatic monitoring of temperature, pH, and possibly acidity of the sourdough
- Filtered air ducts, if required
- Recording of all measured values, metered amounts, and switching operations throughout ripening, whenever possible

Presently available control systems permit the use of sourdough machines not only for single-stage production but for multistage and continuous ripening. The principal objective of such systems is to produce the required quantities of sourdough at exactly the right time [20].

VII. CONTINUOUS SOURDOUGH FERMENTATION

Processes for continuous sourdough fermentation that work with large sourdoughdough ratios, such as the Ankerbrot-Reimelt system and its more sophisticated successor, the Paech-TUB system, have been developed to reduce fermentation time and are used in Germany [1].

A. General Principle

The principle of continuous sourdough fermentation involves pumping a dough through a fermenter in plug-flow pattern as illustrated in Fig. 1. A gradient is created because of changes that occur in the dough between fermenter inlet and outlet, the magnitude of which results from the residence time of the dough in the fermenter. Continuous sourdough fermentation requires the gradient to be constant with respect to microbial count. It must be high enough to permit a adequate acid formation for the intended purpose of the dough.

During sourdough fermentation, the gradient is established by feeding a part of the fermenting dough, taken from the outlet, back into the inlet of the fermenter. This dough serves as an inoculum that is mixed with fresh dough to maintain a constant mass flow. The condition for upholding

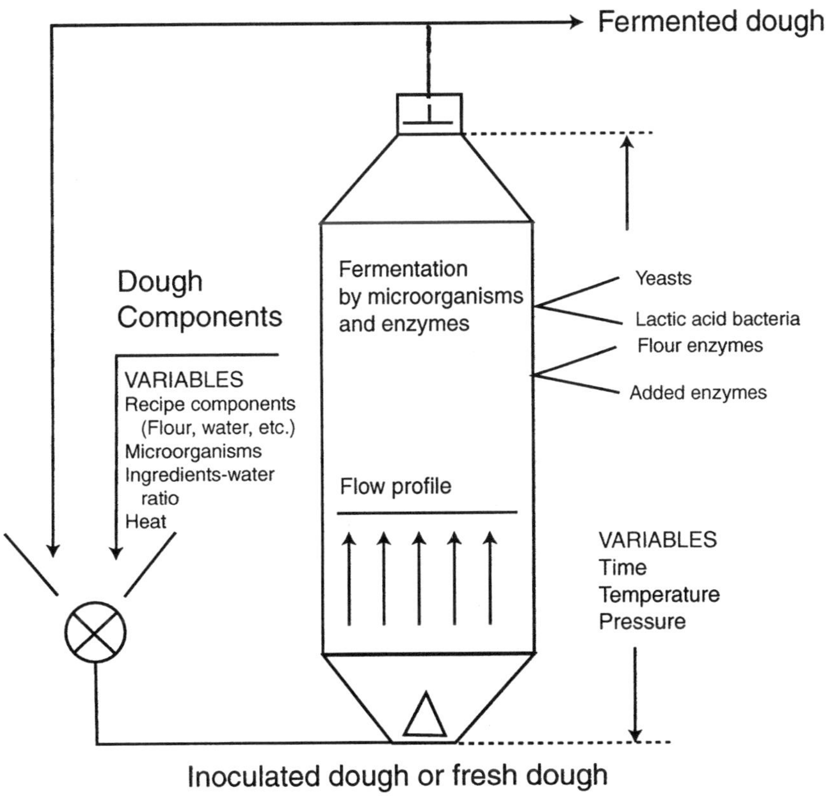

Figure 1 Principle of continuous dough fermentation. (From Ref. [1].)

continuous fermentation requires that, with a constant microbial count at the outlet of the fermenter, the gradient or the ratio between the respective microbial counts at the inlet and outlet of the fermenter remain constant. The residence time of the dough in the fermenter depends on fermentation conditions and, consequently, determines the productivity of the fermenters which is defined as the mass of sourdough that can be taken from the fermenter within a unit of time [1,23,24].

B. Ankerbrot-Reimelt System

The Ankerbrot-Reimelt system (ARS) was the first fermentation system that allowed continuous operation based on the principle of recirculating a portion of sourdough [25,26]. It consists of two fermentation stages, one in a fermentation tube and the other in a fermentation tank. Dough flows from the tube into a mixer to be mixed with flour and water to double the mass flow of the initial dough. Half of the dough is fed to the inlet of the fermentation tube (first stage), while the second half is fed to the inlet of the fermentation tank after its concentration has been further diluted with water (second stage).

This system has some specific advantages. Sourdough can be cooled in the fermentation tube in case of interruptions in bread production. Also only a short time period (less than 4 h) is required to prepare a large mass flow of sourdough for bread production. The preparation time depends on the time needed to generate the required number of microorganisms in the fermentation tube, which is around 3 h under normal operating conditions. The ARS performs well in bakery operations with a large sourdough requirement but with only small variations in production and a small assortment of breads. However, the system cannot handle variations in daily and weekly production demands. Therefore, a new system was needed to meet these requirements such as the Paech-TUB-Reimelt system [1].

C. Paech-TUB-Reimelt System

The Paech-TUB-Reimelt system (PTRS) is based on the principle of combining a continuous and a batch sourdough fermentation to vary production in response to production demands.

The system consists of two fermenters and an insulated tank located between the fermenters, as shown in Fig. 2, connected by two mixers, pumps, and pipes. A heat exchanger is situated after the first fermenter. This fermenter is a narrow, cylindrical, insulated tank through which dough flows continuously. The tank after this fermenter is also cylindrical but has a considerably larger diameter and functions as a storage vessel that is filled

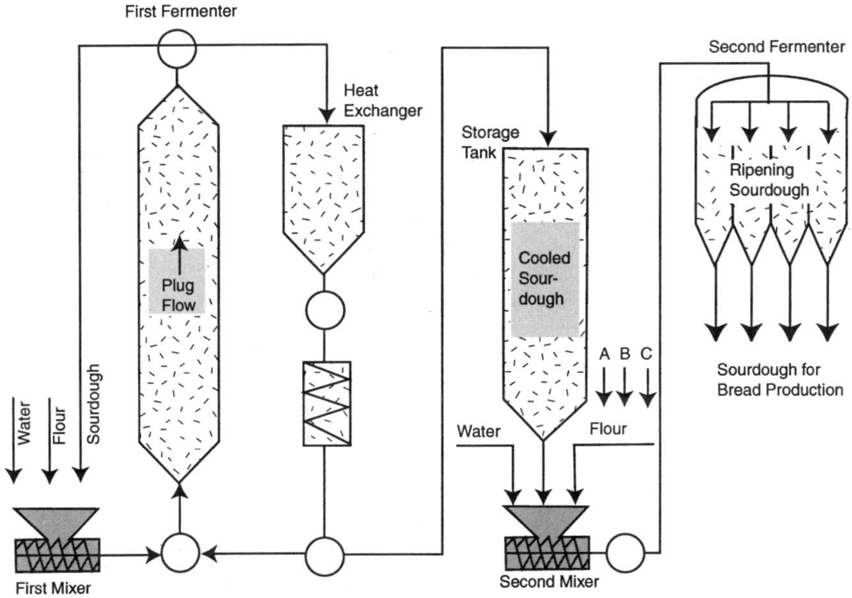

Figure 2 Schematic diagram of the Paech-TUB-Reimelt system used for sourdough production. (From Ref. [1].)

from above and emptied from below. The tank is fitted with a stirrer to facilitate the withdrawal of the cooled dough, which does not flow as easily as freshly fermented dough because of relatively high density and greater viscosity. The second fermenter is constructed as a segmented tank, with the segments also filled from above and emptied from below.

The advantage of this system is that the first stage can be operated continuously over a long period of time. The dough in the storage tank can be used to meet the different requirements for the supply of sourdough. Various quantities can be taken from the storage tank to prepare and propagate fresh sourdough. In the PTRS system, fermentation in the second stage is intermittent because the quantities of sourdough required for each different type of dough are too small for continuous production to be profitable.

VIII. CONCLUSIONS

Although sourdoughs have been prepared for many years, significant progress in this field has been achieved only through the scientific and technical developments of the last two centuries and, in particular, of the last 40 years. The baking industry now uses various fermentation methods as well as many different mixing and dough preparation techniques to produce a variety of bakery products. There is a tendency to simplify and shorten dough-making methods and make them more flexible with regard to the time required for fermentation. Continuous fermentation systems as well as combinations of continuous and batch-processing fermentation systems have been developed for both yeast and sourdough fermentations. The systems can be automated and thus integrated into process computer-controlled production operations in bakeries [1].

REFERENCES

1. Meuser, F. Development of fermentation technology in modern bread factories. Cereal Foods World 40(3):114 (1995).
2. Seibel, W. and Bruemmer, J.-M. The sourdough process for bread in Germany. Cereal Foods World 36(3):299 (1991).
3. Spicher, G. and Stephan, H. Handbuch Sauerteig—Biologie, Biochemie, Technologie, 3rd ed. Behr's Verlag, Hamburg, 1987.
4. Bruemmer, J.-M., Stephan, H. and Spicher, G. Backtechnische Wirkung von Weizenvor- und Weizensauerteigen. Getreide Mehl Brot 38(7):203 (1984).
5. Bruemmer, J.-M. Baeckereitechnologische und analytische Studien an Weizensauerteigen aus Typenmehlen. Getreide Mehl Brot 43(1):19 (1989).
6. Stephan, H. Zweistufige Sauerteigfuehrung. Brot Gebaeck 11(6):122 (1957).
7. Stephan, H. Die Auswirkungen der Fuehrungsweise des Tagesvollsauers auf die Brotqualitaet. Brot Gebaeck 21(12):235 (1967).
8. Lorenz, K. Sourdough processes—Methodology and biochemistry. Bakers Digest 57(4):41 (1983).
9. Stephan H. Detmolder Einstufenfuehrung. Publ. 41. Arbeitsgemeinschaft Getreideforschung e.V., Detmold, Germany, 1960.
10. Stein, vom E. Das neue Monheimer Salz-Sauer Verfahren, Publ. 73 Arbeitsgemeinschaft Getreideforschung e.V., Detmold, 1974.
11. Pelshenke, P. F., Schulz, A. and Stephan, H. Berliner Kurzsauerfuehrung Publ. 72, Arbeitsgemeinschaft Getreideforschung e.V., Detmold, Germany, 1974.
12. Seibel, W., Bruemmer J.-M. and Stephan, H. West German bread. In: Advances in Cereal Science and Technology (Y. Pomeranz, ed.). American Association of Cereal Chemists, St. Paul, MN, 1978, p. 415.
13. Steffen, F.-J. Brotland Deutschland. Deutscher Baecker Verlag, Bochum, 1989, p. 22.
14. Bruemmer, J.-M. and Lorenz, K. European developments in wheat sourdoughs. Cereal Foods World 36(3):321 (1991).

15. Bruemmer, J.-M. Weizensauerteige. 1. Mitteilung: Einfluss der Weizenmehltype, der Teigausbeute, der Temperatur, der Anstellgutmenge und der Reifezeit auf pH- und Saeregrad-Werte. Brot Backwaren 35:302 (1987).
16. Bruemmer, J.-M. Weizensauerteige. 2. Mitteilung: Einfluss von Zusaetzen (Backhefe, Konservierungsstoffe und Weizen-Brot) auf die Saeuerung und Triebleistung. Brot Backwaren 36:370 (1988).
17. Bruemmer, J.-M. Weizensauerteige. 3. Mitteilung: Einfluss der Fuehrungsbedingungen und von Zusaetzen (Backhefe, Konservierungsstoffe, Weizenbrot) auf das Milch—Essigsaeure—Verhaeltnis und die Gasentwicklung. Brot Backwaren 37:78 (1989).
18. Bruemmer, J.-M. Neuere Entwicklungen in der Sauerteigfuehrung und—anwendung. Brot Backwaren 12:298 (1985).
19. Bruemmer, J.-M. and Stephan, H. Einsatz getrockneter Sauerteige und Sauerteig-Starter. Getreide Mehl Brot 40(2):51 (1986).
20. Bruemmer, J.-M. Modern equipment for sourdough production. Cereal Foods World 36(3):305 (1991).
21. Anon. Auf die Fuehrung kommt es an. Baeko-magazin 3:34 (1998).
22. Olms, F. Mechanisierung und Steuerung der Sauerteigbereitung: Iserhaeger Sauerfuehrung. Getreide Mehl Brot 36(2):46 (1982).
23. Vollmar, A. Optimierung der Produktivitaet eines kontinuierlichen Verfahrens zur Sauerteigfermentation unter mikrobiologischen und verfahrenstechnischen Aspekten. Dissertation. Technical University of Berlin, Germany, 1991.
24. Vollmar, A. and Meuser, F. Theoretische Betrachtungen zur Optimierung der Produktivitaet eines kontinuierlichen Sauerteigfermentationssystems. Getreide Mehl Brot 45:331 (1991).
25. Foramitti, A. Kontinuierliche und mechanisierte Herstellung von Sauerteig. Getreide Mehl Brot 36:47 (1982).
26. Dietrich Reimelt KG. Verfahren und Vorrichtung zum kontinuierlichen Herstellen eines fermentierten Teiges. German Patent DE 4125650 (1993).
27. Meuser, F., Zense, T. and Mar, A. Continuous sourdough fermentation with a two-stage pilot fermenter. In: Cereals in a European Context (I. D. Morton, ed.). Ellis Horwwood, Chichester (1987).

11

Sourdough Bread in Finland and Eastern Europe

Teija-Tuula Valjakka and Heikki Kerojoki*
Vaasan & Vaasan Oy, Helsinki, Finland

Kati Katina
VTT Biotechnology, Espoo, Finland

I. BACKGROUND

Sourdough baking is the traditional way of making bread in Finland and northeastern Europe [1]. Most rye bread made in Finland is sourdough rye bread, although in parts of southeastern Finland sourdough bread from barley has also been and is still being made to some extent. Traditionally, sourdough wheat bread has not been produced in Finland. However, wheat preferments fermented with baker's yeast are generally used in wheat baking.

Finns eat 50 kg of bread yearly, of which almost half is rye bread [2]. Rye bread is mostly sourdough rye bread made of whole-grain rye flour. The form and flavor of rye bread varies from region to region. Traditionally, in western Finland rye bread is less sour and is often flat bread, whereas in eastern Finland rye bread is in a form of loaf and is more sour than in the western part. Internationally known specialties are serving size flat breads, thin crisps, and round crisp bread.

---

* *Current affiliation*: UAB Vilnaus Duona Plius, Vilnius, Lithuania.

The most popular rye bread type is small (serving size) flat bread. According to AC Nielsen ScanTrack [2], about half of rye bread consumed in Finland in 2000 was this type of flat bread. Rye loaves are also important. They are consumed especially in eastern Finland, where about half of rye bread in 2000 was in a form of rye loaf.

The consumption of ryebread decreased slightly over 20 years, but during the last 5 years it increased slightly again [3]. Finnish rye bread, unlike other bread types, is made of whole-grain rye flour. This ensures a high dietary fiber content in the final bread. The dietary fiber content of fresh rye bread is approximately 10 g/100 g and approximately 2.5 g per serving (25 g). Rye bread is made without fat or milk, and only some special types of rye bread contain sugar or syrup.

The daily intake of dietary fiber in Finland is approximately 22 g/day/capita. Men's intake of fiber is higher than women's. More than half of the fiber is of cereal origin and as much as 40% of fiber is of rye origin [4]. Thus, from a nutritional point of view, rye bread is also a very important product in Finns' diet.

A lot of research on the health benefits of rye has been carried out in Finland since 1994 in cooperation with the food industry and research institutes. A pioneer in this field is Prof. Herman Adlercreutz, who has studied the benefits of rye in human health during the last 20 years. His studies have concentrated on lignans and phytoestrogens. The lignans matairesinol and secoisolariciresinol are present in the aleurone layer of rye [5]. They are converted in the human body to mammalian lignans by action of the intestinal microbes [6]. Epidemiological, cell culture, and animal experiments have shown that lignans are protective and can lower the risk of prostate and breast cancers [5]. Later, other mammalian lignan precursors have also been found in whole-grain rye [7]. Moreover, according to a consensus statement of experts of the European Cancer Organization, a diet rich in high-fiber cereal is associated with a reduced risk of colorectal cancer, and there is evidence that cereal fiber protects against breast cancer [8]. Recent studies have also shown an inverse association between the intake of dietary fiber and the risk of coronary heart disease [9].

Rye, despite being internationally a marginal crop, is becoming increasingly popular in Finland. Both baking companies and organizations have given information to consumers about the health benefits of whole-grain rye bread.

II. SOURDOUGH RYE BREAD IN FINLAND

A. Preparation of Rye Sourdough

Typical Finnish rye sourdough is made of whole-grain rye flour, water, and starter. The starter is usually a portion of sourdough from a previous sourdough batch. Pure starter cultures or baker's yeast are not generally used. In an earlier time, sourdough was made at home and in bakeries in wooden bowls, which were not washed after use. Thus, the remains of sourdough dried to the edges of the bowl and functioned as a starter for the next sourdough batch. The microbiology of Finnish sourdoughs has been studied by Salovaara and coworkers [10]. A more detailed description about the characteristics of sours is given in Sec. II.B.1.

Each bakery has its own sourdough starter. This starter culture might be decades old and is taken care of with great attention. However, we have to remember that the microflora of whole-grain rye flour, possible changes in fermentation temperature, and water/flour ratio of sourdough undoubtedly have an effect on sourdough flora. So, despite great attention paid to the microflora of bakeries' sourdough, starter probably changes over time.

On weekdays, when sourdough is used in production, the sourdough starter is renewed daily. On weekends, bakeries have a 1- to 2-day break in production, which requires special attention to the sourdough. The most common way is to keep a sourdough starter in the refrigerator on weekends. At the lower temperature fermentation slows down but the microflora stays alive. When sourdough is taken back to fermentation temperature, fermentation continues and a new sourdough batch can be produced.

Another way to store sourdough starter is to add flour to the sourdough to decrease water activity and to slow down fermentation. This kind of sourdough cake has to be stored in the refrigerator. Hurri-Martikainen [11] added sterilized and regular whole-grain rye flour to sourdough. The amount of added flour was 50% of the sourdough amount.

Table 1 Typical Formula of Single-Stage Rye Sourdough

Factor	Typical value	Range
Sourdough starter	1.5 kg	1.5–5 kg
Whole-grain ryeflour	60 kg	50–80 kg
Water	100 kg	100 kg
Fermentation time	16 h	12–24 h
Fermentation temp.	28 °C	25–32 °C
Sourdough yield	267	220–300

The fermenting and leavening capacity as well as the yeast count decreased during storage. After 4 weeks storage leavening capacity was only 50% of fresh sourdough's leavening capacity. Typically, the amount of added flour is much higher at 100–150% of the sourdough amount. Sourdough cake is usually used as a starter for new sourdough within a month. Although some bakeries use this kind of method successfully, it is not clear how well sourdough yeasts and bacteria preserve in this kind of somewhat dry sourdough cake.

Temperature control is critical in sourdough production. Changes in fermentation temperature may cause variation in microflora of sourdough and thus variation in sourdough and final bread quality and flavor.

Earlier sourdough was fermented overnight in open containers at bakery ambient temperature. This is also the basis for present sourdoughs. Commonly, sourdough is fermented for 12–18 h at 25–32 °C. Also colder and warmer fermentation temperatures are used. During fermentation lactic and acetic acids are formed. The amount of these two acids has great impact on bread flavor. With a higher proportion of lactic acid bread flavor is mild, whereas acetic acid gives a stronger flavor to the product. Thus, a certain proportion of acetic acid is often desired [10]. If acetic acid and yeast growth are preferred, fermentation temperature is lower, e.g., 20–25 °C. On the other hand, if lactic acid production or milder acid flavor is desired, fermentation temperatures as high as 32–38 °C can be used. It is known that homofermentative lactic acid bacteria favor higher fermentation temperatures. Thus, if a higher proportion of lactic acid is preferred, higher fermentation temperatures should be used.

The water/flour ratio of sourdough also varies from bakery to bakery. Often the sourdough yield is 250–300 [10]. It is a practical sourdough consistency. This sourdough is easy to make and pumpable, but not too watery. A typical formula of one stage rye sourdough is shown in Table 1. Chemical characteristics of Finnish sourdoughs are described in Sec. II. B.3.

Today most bakeries use the one-stage sourdough process. Earlier two- or multiple-stage processes were commonly used to standardize the quality and quantity of sourdough. In the two-stage sourdough process, sourdough is made as in the one-stage process, but it is refreshed before use. In refreshening, a certain portion of flour and water is added to the sourdough and it is fermented for a couple of additional hours [12]. Today the multiple-stage process is used if yeast growth or special aroma is desired. Formula and conditions for multiple-stage rye sourdough is shown in Table 2.

Rye flour used in sourdough is whole-grain rye flour. Whole-grain flour (ash content 1.8% d.m.) is used for its higher buffering capacity and hence greater acid forming properties. If sieved flour is used, pH drops more

Table 2 Traditional Multiple-Stage Rye Sourdough

Stage	Quantity
Stage 1	
Sourdough starter	3 kg
Whole-grain rye flour	30 kg
Water	39 L
Fermentation time	Overnight
Fermentation temp.	25–26 °C
Sourdough yield	230
pH of fermented one-stage sourdough	3.8–3.9
Stage 2	
One-stage sourdough	24 kg
Whole-grain rye flour	20 kg
Water	21 l
Fermentation temp.	28 °C
Fermentation time	2.5–3 h
Sourdough yield	214
pH of ready two-stage sourdough	3.9–4.2

Source: Ref. 12.

quickly to a value at which acid formation of sourdough bacteria gets weak. Thus, sieved rye flour produces lower acidity than whole grain rye flour [13]. The flavor of sourdough made of whole-grain rye flour is more pleasant and stronger than sourdough made of sieved rye flour (ash content less than 1.2% d.m.).

In sourdough low falling number whole-grain flour (falling number 70–100) is preferred. During sourdough fermentation the forming acids inactivate excess enzyme activity; therefore, very low falling number can also be used in sourdough making. When low falling number rye flour is used, fermentation is faster and the flavor of sourdough is more pleasant. Although low falling number rye flour is preferred, regular whole-grain rye flour (falling number 120–140) can be and is used by most small and medium-sized bakeries. Generally, rather coarse flour is used. Even coarser meal-type flour can be used as well, but rarely is sourdough made of fine rye flour. The specification for sourdough rye flour is shown in Table 3.

Today larger bakeries have temperature-controlled sourdough containers (Fig. 1). Sourdough is made in a container having two to four mixing blades in its axle. These containers can be provided with an automatic mixing schedule and water collar to regulate the fermentation temperature.

Table 3 Specification of Whole-Grain Rye Flour for Sourdough

Measurement	Whole-grain rye flour for sourdough	Whole-grain rye flour for dough	Method
Ash content	1.8 ± 0.15% d.m.	1.8 ± 0.15% d.m.	AACC 08–01
Falling number	85 ± 15	120 ± 15	ICC Stand and No. 101
Moisture content	Max. 14%	Max. 14%	ICC Standard No. 110/1
Sieve analysis:			
>670 µm	15–20%	15–20%	
>125 µm	40–65%	40–65%	
>100 µm	5–10%	5–10%	

Source: Melia Oy, Raisio, Finland.

Figure 1 Sourdough tank. (Courtesy of Ipeka Automation Oy.)

Large bakeries have also automatic metering systems for flour and water. The typical size of a sourdough tank in a larger bakery is 4000 L. Sourdough increases in volume during fermentation; thus, sourdough tanks are filled maximally to 75% of their volume. When making a sourdough batch, the temperature of the water collar is adjusted to the desired temperature and water is added to the sourdough tank manually or automatically. Sourdough starter is added manually through a manhole on top of the tank. The mixer is started and during mixing the axle changes its direction of rotation according to a programmed schedule. The needed amount of flour is added manually or automatically. During fermentation sourdough is mixed at long intervals to keep the sourdough homogeneous. The tank is washed after use.

B. Characteristics of Sours

1. Microbiological Characteristics

Finnish sourdough is generally made of whole-meal flour, water, and starter. Whole-grain rye flour (extraction rate 100%) may contain unspecified bacteria between 10^4 and 10^6 cfu/g. Counts of coliform bacteria are 10^2–10^4 cfu/g and for spontaneous flora of lactic acid bacteria (LAB) counts are 10^2–10^3 cfu/g. With a typical Finnish sourdough process with a starter from a previous batch, the final counts of lactic acid bacteria at the beginning of fermentation are 10^7–10^8 cfu/g. In a fully fermented rye sourdough there are around 10^9 lactic acid bacteria per gram. The number of yeast cells is 10^7–10^8 cfu/g [14].

Most Finnish sourdough systems are quite stable in spite of repeated contamination by microorganisms in flour, indicating the presence of a well-adapted natural flora. Both homofermentative and heterofermentative strains of LAB have been found in sourdoughs. The most frequently identified LAB in Finnish sourdoughs are presented in Table 4. Homofermentative LAB seem to be the dominating species in the sourdoughs; the most frequently identified LAB in Finnish sourdoughs is *Lactobacillus plantarum*. Two main types of yeast, *Saccharomyces cerevisiae* and *Candida milleri*, have been identified in Finnish sourdoughs (Table 4). *C. milleri* is a frequently identified yeast in industrial sourdoughs.

2. Chemical Characteristics

Usually, the rye sourdough process involves both lactic acid and yeast fermentation. Due to these fermentations, acidity of sourdough will increase. Mature sourdoughs have a pH value of 3.5–5.4 and total titratable acidity (TTA) values of 9–18. The lactic acid content of sourdough is

Table 4 Typical Lactic Acid Bacteria and Yeasts Identified in Finnish Sourdoughs

Species	Ref.
LAB	
Lb. acidophilus	[41]
Lb. fructivorans	[45]
Lb. casei	[41]
Lb. sanfransico,	[44, 45]
Lb. plantarum	[41, 43, 44, 45]
Lb. viridescens	[41, 44]
Lb. bucheri	[41]
Lb. cellobiosus	[41]
Lb. brevis	[44]
Yeast	
S. cerevisiae	[42]
C. milleri	[42, 44]

typically 0.44–0.93% and content of acetic acid is 0.03–0.019% [11]. Development of acidity and microbial counts in sourdough are presented in Figs. 2 and 3. The amount of acids formed is dependent on the composition of microbial flora of sourdough and fermentation parameters such as time and temperature. Homofermentative LAB produce mainly lactic and heterofermentative LAB produce both lactic and acetic acids, carbon dioxide, and ethanol. The pH drop is mainly caused by lactic acid formed. Acetic acid has a remarkable role in sensory acidity and in enhanced microbiological shelf life. Heterofermentative LAB may also have a role in raising the dough if sourdough does not contain yeast [15].

C. Uses of Sourdoughs

1. Sourdough Rye Bread

Finland has a very rich bread tradition. In western Finland bread preferences had influences from Sweden and in eastern Finland from Russia. All over Finland rye bread has traditionally been sourdough bread. In western and northern Finland bread has not been as sour as in eastern Finland but nevertheless has been made with sourdough fermentation. In western Finland bread is often also slightly sweet. Especially in south-

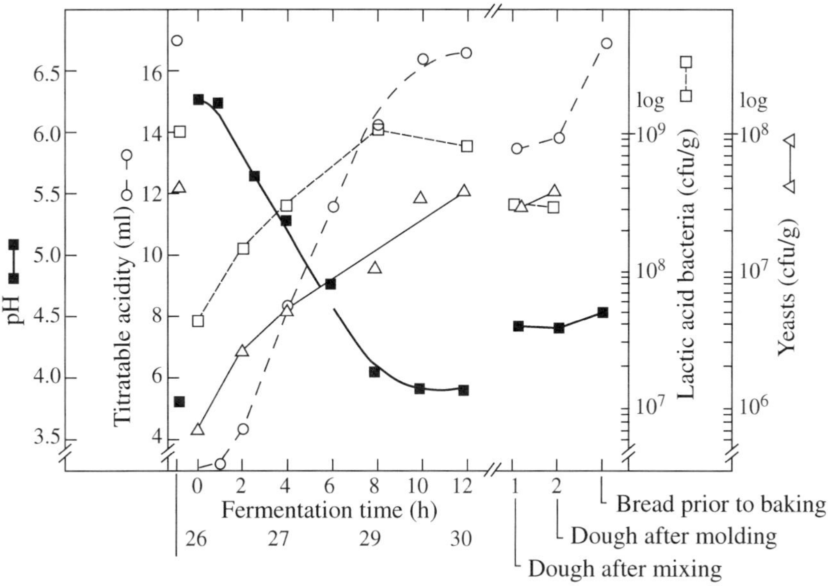

Figure 2 Development of acidity and microbial counts in industrial rye sourdoughs [14].

western parts of Finland, spices such as caraway are used in sourdough bread.

The form of bread has traditionally been different in different parts of Finland. In eastern Finland bread has been baked in the form of a loaf (Fig. 4), and in western Finland a special type of round flat bread with a hole in the center has been very common (Fig. 5). Nowadays flat breads in various forms and sizes are increasingly popular throughout Finland.

Typical rye dough is made of sourdough, water, whole-grain rye flour, wheat flour, yeast, and salt. Traditionally, the dough liquid came exclusively from sourdough, but nowadays water is often added. A typical rye bread formula is shown in Table 5. Most rye bread is made of whole-grain rye flour. In some parts of Finland light rye flour (ash content 0.9% d.m.) replaces part of the rye flour. Rye bread can be made exclusively of rye flour, but wheat flour is often used to improve dough handling properties. Dough made with rye flour is sticky and difficult to handle. The amount of wheat flour is usually less than 25% of total flour [16]. Wheat flour can be white wheat flour (ash content 0.7% d.m.) or whole wheat flour (ash content 1.5% d.m.). Besides improving dough handling properties, wheat flour also improves bread crumb grain and texture. The wheat flour ratio also affects bread flavor.

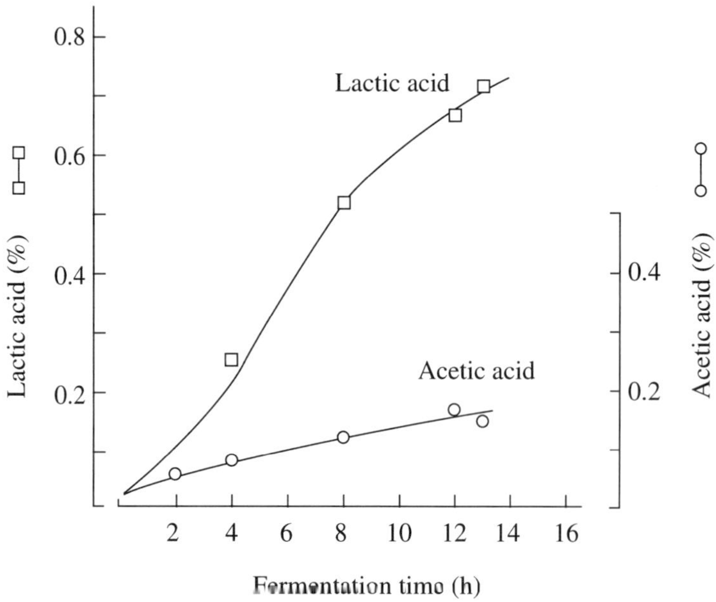

Figure 3 Development of lactic acid and acetic acid during Finnish rye sourdough fermentation. [From Ref. 10, with permission by Hannu Salovaara.]

The amount of sourdough-fermented flour is generally 10–50% of the total flour [16]. Sourdough quality has an effect on dough fermentation and thus on loaf volume and bread crumb structure. Obviously, the quality and quantity of sourdough has a major impact on bread flavor.

Most bakeries add baker's yeast to bread dough. If baker's yeast is not used, the sourdough process is adjusted so that the yeast growth is preferred. Bread dough is leavened by the action of yeasts and LAB in sourdough. However, bread baked without baker's yeast is denser than regular sourdough rye bread containing some baker's yeast.

Traditional sourdough loaves are still made in great number especially in eastern and central Finland. A typical rye loaf formula is shown in Table 5. Rye dough does not need intensive mixing as a wheat dough. Usual mixing time is 5–10 min depending on formula and type of mixer. Dough is fermented about an hour before scaling. Typical loaf size is 500–800 g. Dough is pressed though a special extruder and cut to the desired weight by knife or thread. Weighted pieces are then rounded by hand or mechanically with a special type of rounder having two kneading boards or two belts moving at different speeds. Rounding is not possible with a rounder

Figure 4 Rye loaf. (Courtesy of Vaasan & Vaasan Oy.)

Figure 5 Rye flat bread. (Courtesy of Vaasan & Vaasan Oy.)

designed for wheat doughs. After rounding dough is proofed and baked as described in Table 5. Rye bread is proofed under rather dry conditions. A slightly floury, cracked crust is desirable especially in a rye loaf, as shown in Fig. 4.

Another typical sourdough rye bread is flat bread. The traditional form is round (diameter approximately 28 cm) with a hole in a center. Today various sizes and forms are used. Serving size flat breads are most popular today.

Flat bread dough has a higher dough yield than rye loaf dough and does not need as long a fermentation as rye loaf dough, as seen in Table 5. Wheat flour is also added to flat bread dough to improve dough handling. Today flat breads containing rye grits or pieces of rye kernel have become increasingly popular.

Flat bread dough is soft and sticky and thus difficult to handle manually. Flat bread is formed with a special machine called the "Kantti-Expert (LT-Kone Oy, Finland)." Dough is handled on a conveyor line. The conveyor belt is dusted with rye flour to prevent the dough's sticking on the belt. The dough is put into the dough hopper and extruded by twin rolls to form a dough mat of uniform thickness. The top side of the dough is dusted with rye flour, docked and cut to the desired form. Cutting is done with a stamping mold. After cutting the left-over dough is mixed with fresh dough in the dough hopper.

Table 5 Typical Rye Bread Formula

Factor	Flat bread	Loaf
Sourdough	260 kg	100–150 kg
Whole-grain rye flour	100 kg	100 kg
Wheat flour	0–30 kg	0–30 kg
Salt	3–4 kg	3–5 kg
Yeast	3.5–5 kg	3–5 kg
Dough temp.	30 °C	30 °C
Dough fermentation	0–30 min	30–60 min
Dough weight per piece	400 g	500–1000 g
Dough yield	180–190	150–160
Amount of fermented flour	40–46%	25–30%
Proofing time	30–60 min	50–60 min
Proofing temp.	36–38 °C	30–35 °C
Relative humidity in proofing	65–70%	65–70%
Baking time	20–30 min	45–75 min
Baking temp.	250–280–210 °C	250–280–210 °C

Larger bakeries have automated their bread lines. Dough pieces are transferred automatically from dough made up to a cabinet-type proofer. From the proofer dough pieces are automatically transferred to oven. Typical proofing and baking times and temperatures are shown in Table 5. After baking breads are cooled for approximately 30 min.

Flat rye breads have a moist crumb. The proportion of crust is rather high, which gives a lot of flavor to the product [16]. Today most flat breads are cut in two pieces after baking and cooling (Fig. 6). Cutting is carried out with a special machine in which bread is forced through a moving knife, which splits the bread. The bread is sliced in a similar way as a hamburger bun. Development of the cutting technique has changed the way of using rye bread in Finland to a remarkable degree.

A special type of sourdough rye bread is called rest-heat bread. At an earlier time when wood-heated baking ovens were used, this type of bread was the last product to be baked and so it was baked at a lower temperature (rest heat) for a longer time. Rest-heat bread has a thick crust and very strong flavor. Because of the longer baking time the crumb is drier and darker and the flavor is slightly sweet, although sugar is not used in the formula. Nowadays most rest-heat breads are flat breads, but loaf-type

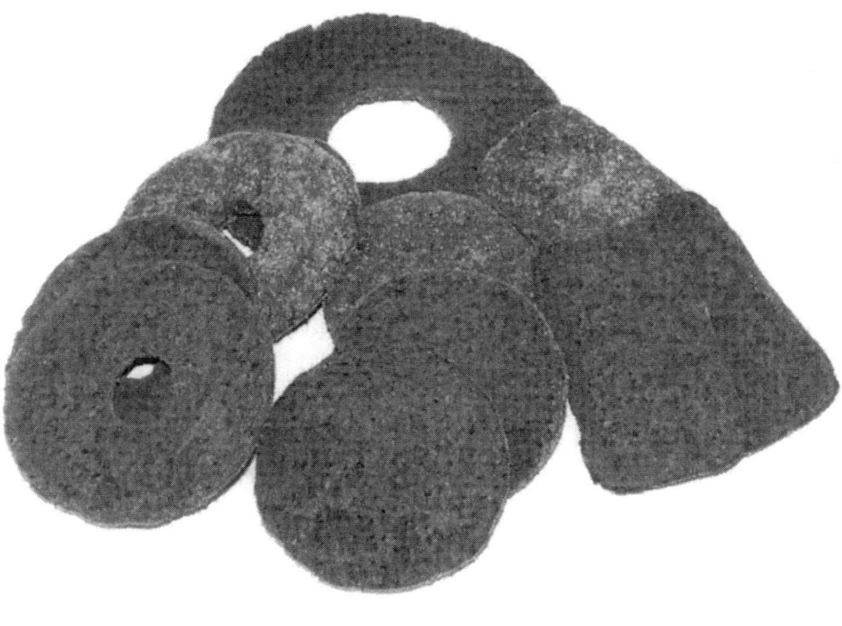

Figure 6 Cut rye flat bread. (Courtesy of Vaasan & Vaasan Oy.)

bread can also be baked with this process. Bakeries have developed their processes to produce this type of rye bread.

Most rye breads are sold packed in polyethylene bags. The keeping properties of rye bread are good. The shelf life is about one week without preservatives or other additives. Cut or sliced rye breads are more susceptible to mold growth. The flavor of rye bread does not deteriorate as quickly as that of wheat bread. Shelf life of cut flat bread without the addition of preservative is generally 4 days.

2. Sourdough Crisp Bread

In the 19th century, especially in western Finland, round sourdough rye bread was dried on sticks hanging from ceilings of farm houses. This way bread kept its eating properties for a long time [17]. Similar dried rye bread is still produced by some bakeries. It is popular in western and central–western Finland. The bread has a similar formula and process as that of soft flat bread, but after baking the bread is cut and the bread halves are dried at moderate oven temperature for 30 min. Dried rye bread has a strong rye flavor and a rather hard texture, and requires a lot of chewing. According to AC Nielsen Scan Track [2], approximately 10% of crisp breads sold in Finland are dried rye bread (Fig. 7).

Industrial production of crisp bread started in Finland in the beginning of the 20th century. Although traditional crisp bread is a rye product, it is not necessarily a sourdough product. Very thin and crispy crisp bread, called ice bread, is produced without dough fermentation [18]. This type of crisp bread is most common in central Europe and Great Britain. In Scandinavia crisp bread is most often yeast fermented and especially in Finland sourdough crisps are common.

Even though crisp bread is considered a Swedish product, thin sourdough crisps are originally from Finland and are produced only in Finland. The product is very thin, with an approximate height per piece of 3 mm. Most thin crisps are cut to serving size pieces (Fig. 8). They are rather sour having a pH of approximately 4.5 and a total titratable acidity of 20–25. Moisture content is less than 5%.

Thin rye crisp dough is somewhat similar to flat rye bread dough. Dough ingredients are sourdough, whole-grain rye flour, water, yeast, and salt. Wheat flour is not generally used. Products can be spiced with caraway seeds. Fat or sugar is not added to rye crisp bread dough. Some crisp breads contain emulsifier. Dough is fermented for 1–2 h before sheeting. Because the sheeted dough has to be very thin, it requires a soft, uniform, and extensible dough. Before sheeting, the dough is remixed to obtain a uniform dough structure. Dough is then sheeted to even thickness of approximately

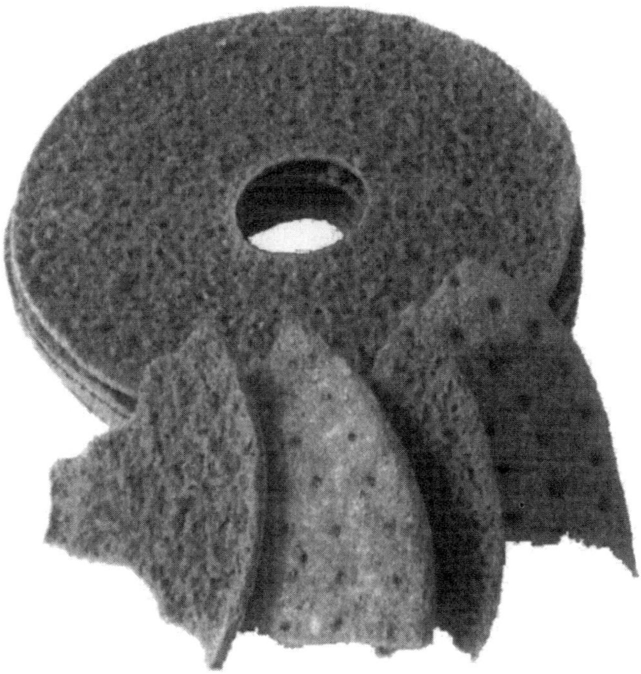

Figure 7 Dried rye bread. (Courtesy of Vaasan & Vaasan Oy.)

3 mm with a single pair of forming rolls. The sheeted dough is dusted, docked, and cut to the desired size. Cutting is done with knives or a stamping mold. The dough is then proofed for 20–40 min at approximately 30 °C and thereafter baked and dried.

Baking time is short, usually less than 10 min at rather high temperature. The controlled baking procedure ensures the right texture characteristics. The water activity in crisp bread is lower than 0.7 and the moisture content less than 10%, which makes crisp bread unsusceptible to microbiological contamination. The shelf life of crisp bread is at least a year, if the product is stored in a dry place and protected from light.

D. Effects of Sourdough

1. On Dough

Rye has a clearly inferior baking quality in comparison with wheat [19]. Rye dough is often sticky and the texture of the bread is dense. This is partly due

Figure 8 Sourdough rye crisps. (Courtesy of Vaasan & Vaasan Oy.)

to lack of a gluten network of rye and partly due to the higher solubility of rye proteins. Thus, proteins have a minor role in rye dough. Pentosans, however, are believed to have a more important role in dough rheology and gas holding properties in rye dough. Pentosans constitute the major water-binding substance in rye dough and partially determine the rheological properties of rye dough. They also regulate water distribution in dough during baking, affecting starch properties such as gelatinization and retrogradation. Pentosans also slow down diffusion of carbon dioxide from wheat dough and resemble gluten in this respect [20,21].

In Finland, rye bread is baked from whole meal containing bran and coarse endosperm particles, which affects the rheological properties of rye dough. Large particles partly determine the stiffness of dough, and if the amount of these particles increases, the stiffness of dough increases [22]. The degradation rate of cell walls and the swelling of cell walls affect rheological properties of dough. In rye dough, very rigid particles are dispersed in a weak continuous phase composed of protein, starch, and cell wall components [22,23]. In wheat dough, the continuous phase (proteins and pentosans) is responsible for rheological properties.

In rye baking, starch, proteins, and pentosans absorb water and swell during mixing. Rye has more soluble components than wheat flour. These components have major effects on rye baking quality. At the beginning of the rye baking process, cell walls and flour particles are more or less intact, but during the baking process, cell walls start to degrade due to enzymatic activity of flour. The amount of water and the falling number of rye flour affect the degradation process of cell walls. Endogenous enzymes of rye flour (amylases, pentosanases, and proteases) affect solubility and swelling of flour components. Enzyme activity is strongly dependent on pH; thus, control of pH in rye baking is very important.

Most Finnish rye breads are acidic since they are made by using sourdough. Acidity improves the baking quality of rye, which is partly based on improved swelling and degradation of pentosans and proteins during fermentation. Acidity also restricts excessive activity of amylases in rye dough.

The enzymatic activity of rye dough usually increases until the total titratable acidity value is 10–12. As dough acidity increases further, enzyme activity diminishes preventing excessive degradation of starch and proteins [24]. Sourdough modifies the structure of rye starch during fermentation, which increases the water absorption capacity of starch [25]. Low pH also increase the solubility and swelling capacity of pentosans; optimal solubility and swelling are obtained at pH 4.9 [26].

Historically, the leavening capacity has been the critical property of sourdough, but today this function is not of major significance because of the low price of compressed baker's yeast or active dry yeast. However, some Finnish bakeries still have rye bread products on the market that are manufactured by using only sourdough yeasts for raising the product. Leavening of dough is normally based on alcoholic fermentation by yeast. Yeast strains of sourdoughs differ in their ability to produce CO_2; *C. milleri* has been shown to be the most effective gas producer [27]. In rye sourdoughs the heterolactic fermentation by heterofermentative LAB also has a role in gas production as, for example, *L. brevis* ssp. *linderi* has been postulated to significantly contribute to dough leavening in rye sourdough [28]. Optimal gas production demands careful control of fermentation conditions in the rye sourdough process.

2. On Finished Product

Flavor. The sourdough process is essential for original rye bread flavor as chemically acidified rye bread has inferior flavor [29]. Due to lactic acid fermentation, lactic acid, acetic acid, and other volatile compounds, mainly carbonyls, are produced in various amounts depending on the microflora of sourdough and fermentation conditions [30].

Homofermentative and heterofermentative LAB produce clearly different flavors in bread. Heterofermentative LAB produce strong spicy, acidic, and strong crust flavors, whereas homofermentative LAB produce softer, sweeter, more "flower-like" flavor [31]. The ratio of lactic acid to acetic acid is especially important for flavor. In Finnish rye bread it is usually around 80:20. Typical pH values in Finnish rye bread are 4.4–4.6 and TTA values are 10–14 [11]. Flavor is flat and nonaromatic if the concentration of lactic acid is too high, and if the concentration of acetic acid is too high, flavor is too acidic [32]. Proteolysis during sourdough fermentation also has a role in bread flavor as free amino acids are potential flavor precursors for bread crust flavor, which is formed during the baking stage mainly due to Maillard reactions [33].

Research on the sensory quality of Finnish rye bread is limited. According to Hellemann et al. [34], the most salient sensory attribute in Finnish rye bread is perceived sourness. The importance of sourness to the flavor of sour rye bread has been demonstrated in various studies [35,36]. According to Heiniö [37], the acidity of rye bread and the ash content of rye flour used in the sourdough have the greatest effect on the identity of rye bread.

Antimicrobiological Properties. Microbial spoilage is not usually a very serious problem with rye bread due to the acidic nature of this bread type. However, some sliced rye breads may suffer molding from time to time. The most common contaminating stages are cooling, slicing, and packaging of rye bread. Contamination originates usually from flour dust in the bakery. The most common spoilage molds belong to *Aspergillus* and *Penicillium*.

Sourdough is thus very effective in enhancing microbiological shelf life in rye products. The efficacy of sourdough is based on lactic and acetic acids, and other potential antimicrobial compounds formed during the fermentation. Acetic acid especially is shown to be very effective in mold and rope prevention. The inhibitory effects of sourdough are assumed to be the result of a combination of acids, low pH, and other antimicrobial substances present in the sourdough [38,39]. The low pH favors the undissociated form of weak acids, which was shown to be responsible for the antimicrobial effect. The possible antimicrobial metabolites in sourdough might be bacteriocins or low molecular mass compounds with a wide spectrum of activity against gram-positive and gram-negative bacteria and fungi.

III. SOURDOUGH RYE BREAD IN BALTIC COUNTRIES

A. General

Rye consumption in Baltic states is traditionally at a very high level: Estonia 19.9 kg, Latvia 14.7 kg, and Lithuania 52.3 kg per annum per capita. Lithuanian rye consumption is the highest in the world. The basic bread in the Baltic republics' everyday diet is often a wheat bread loaf, which is generally not regarded as a valuable part of a diet. Rye bread is regarded as a high-value part of a healthy diet. Rye bread is more consumed in rural areas but is popular in cities as well. The consumption of rye bread is regarded to be stable in all three Baltic republics.

Rye is used for rye bread, or so-called sweet sour bread, called Saldskabmaize in Latvia. Sweet sour bread could be made from wheat also, but is mainly made from rye with a maximum 20% of wheat flour.

Traditionally, country rye bread loaves were often very big. Even today in Latvia one can buy a rye loaf weighing 5 kg. Today these breads are still produced with traditional methods using much hand work, and they may be baked in a big stone oven. For the modern consumer's convenience, these huge loaves are available sliced and packaged. In Lithuania, it is still common to buy an unsliced 2-kg loaf, packaged in a colorful plastic bag.

B. Raw Materials

Estonia, Latvia, and Lithuania have their own rich tradition of making rye bread. The Baltic rye breads are traditionally based on cooked rye flour, which gives the product a sweetish taste. Sterilizing the rye flour by cooking ensures a stable quality of substrate for sourdough fermentation. Cooking also eliminates the potential problem of sprouted rye and too high enzymatic activity.

Today cooking of flour is not always done. Sweetness of the end products is obtained by adding syrup (10–15% of dough weight) at the dough stage to give the product the desired sweet taste. Often syrup or sugar is added to the dough when the traditional flour cooking method is used. Rye and sweet sour breads should be slightly sour, but they are supposed to have a sweet note in taste, especially in the case of sweet sour bread.

Malted rye flour, enzymatically active or nonactive, is generally used in rye and rye-based sweet sour bread formulations. Enzymatically active rye malt helps the gelatinized starch to split into dextrins and sugars.

Caraway seeds are often added to doughs, often in the sourdough process. Caraway seeds are an essential part of rye bread and sweet sour bread in the Baltic region. The seeds are mostly used whole. Some products may be even decorated with caraway seeds.

Baker's yeast is not necessarily used in traditional rye bread. The gassing power is formed in sourdough and in dough fermentation. The long process times support the formation of enough yeasts to take care of the dough leavening. Table 6 presents formulations used in the Baltic republics.

C. Processes

1. Traditional Process

Traditionally, rye flour was cooked, or, to be precise, boiling water was poured onto rye flour in a special wooden barrel (Latvian name "Abra"). It was vigorously stirred with a wooden paddle to produce uniformity. The

Table 6 Typical Rye Bread Formulas (Kg)

	Traditional rye bread	Soviet "brick" bread	Traditional sweet sour bread
Flour cooking (made 24 h before batchwise):			
Rye flour			
(ash 1.42% d.m.)			23
Water (+95 °C)			39
Rye malt flour			3.0
Caraway seeds			1
Sourdough type:	Liquid	Stiff	Liquid
Rye flour			
(ash 1.42% d.m.)	20	33	4
Rye Malt flour	5	—	0.7
Caraway seeds	1	—	0.1
Water	65	25	9
Dough:			
Sourdough (see above)	91	58	79.8
Rye flour (ash 1.42% d.m.)	65	37	42
Wheat flour (ash 0.74% d.m.)	10	—	27
Wheat flour (ash 1.2% d.m.)	—	30	—
Salt	1	1.5	1.5
Sugar Syrup	10	5.	3
Skimmed dry milk	(0.5)	—	(0.7)
Baker's yeast	(0.2)	0.5	(0.2)
Water	—	50.8	—

Parentheses mean's optional; —, means: does not apply.

"porridge" was allowed to cool with occasional stirring. During the cooling period, which could last 24 h, the gelatinized starch partially hydrolyzes into dextrins and lower molecular weight sugars to give the desired sweetness.

The sourdough starter was kept in a wooden barrel and added to the cooked flour the next day to continue the sourdough process. Finally, the rest of flour was added and the final dough was mixed. After a couple of hours of ripening the loaves were formed by hand in a long oval shape. Decorations, if used, were cut on the sides of the breads.

Big properly heated stone ovens have been used to bake traditional rye and sweet sour breads. The baking temperature has been relatively high in the beginning, up to 300 °C. The bigger the loaf, the longer the baking time needed. The 5-kg loaf requires 3 h to be baked properly. The color of bread should be quite dark, nearly black, but not charred. The crust of the traditional loaf is very thick, often more than 1 cm. After baking, the breads were glazed with a starch slurry to create an appetizing shine. The processes for the production of these breads are outlined in Table 7.

Traditionally baked rye and sweet sour bread has a very long shelf life. Two weeks is not exceptional.

2. The Soviet Tradition

During the years the Baltic countries were under Soviet rule, Moscow controlled bread production in a standardized way according to the All-Union standards. Bread was the essential part of the diet in the Soviet Union, and it was strictly kept always available. The bakeries built under the Central Plan were designed to assure a much higher production that was actually needed. Due to subsidized prices, bread was very cheap and available, and was often fed to home animals when appropriate feedstuff was not available or if there was a deficit of feed.

Bread research concerning sourdough and rye bread was at a high level in the Soviet Union. The Moscow Bread Institute had a skilled scientific staff, including technicians, technologists, microbiologists, and dietitians. Many sourdough processes were based on pure cultures of microorganisms, thus stabilizing the quality of bread. Today many of these Soviet processes still are in use, and a staple bread of excellent quality is produced using these technologies.

The mechanized way of baking Soviet-type rye and rye-wheat breads is carefully controlled. The most popular bread has been the so-called brick. It is still among the favorite breads today. The brick is in a form-baked 600- to 700-g loaf that contains 60% (50–80%) rye flour (ash content 1.42% d.m.) and 40% (20–50%) wheat flour (ash content 1.2% d.m.). It is made of very wet dough (water content about 50%) with special technology. The final

Table 7 Typical Rye Bread Process

Factor	Traditional rye bread	Soviet "brick" bread	Traditional sweet sour bread
Dough temp.	30 °C	29 °C	33 °C
Dough fermentation	120 min	70 min	180 min
Dough weight per piece	755	695	710
Dough yield	177	183	154
Amount of fermented flour (rye malt flour included)	25%	33%	30%
Proofing time	55 min	60 min	60 min
Proofing temp.	35 °C	35 °C	35 °C
Relative humidity in proofing	85% R.H.	75% R.H.	85% R.H.
Baking time	47 min	41 min	60 min
Baking temp	320 °C (I zone) 215 °C (II–IV zones)	280 °C	350 °C (Prebaking zone)—280 °C–200 °C (End)
Bread acidity, total titratable acidity	8–10	8–10	6–7
Bread weight (baking loss)	655 g (13.2%)	605 g (12.9%)	611 g (13.9%)

baked product has an even and fine-pore structure and, thanks to the high water content, a relatively long shelf life.

Baking of traditional rye and sweet sour bread is done according to the old traditions. Modern techniques are utilized, often based on the inventions of the Moscow Bread Institute.

3. Preparation of Rye Sourdough

Depending on the bakery, the sourdough is either liquid (pumpable; dough yield 300) or stiff (dough yield 170). There are many alternative variations to control the process regarding the stability of the sourdough. Here we concentrate on the two main types: liquid and stiff. Preparation of liquid sourdough in presented in Table 8.

Liquid Sourdough. Liquid sourdough is made using cooked (gelatinized) rye flour. The rye flour type may vary from R1000 (ash content 1.0% d.m.) to R1800 (ash content up to 1.9%). Most common is the type with ash content of 1.42% (on d.m.), R1450 (ruzdu skrotetie milti in Latvian). The flour type used often depends on availability (Table 9).

The processes have been designed to allow a wide variation of possible raw material quality and still ensure products of stable quality. Flour is typically cooked (gelatinized) in horizontal cookers equipped with effective agitation. Water (67%) is boiled, or heated, up to +90 °C and flour (33%) is added simultaneously with the water. After 50–70 min agitation the obtained smooth slurry is pumped to cooling vessels (bins). It is allowed to cool to a temperature required by the next process stage. The cooling vessel may be equipped with a water jacket for faster cooling but is often without. The cooling time is typically up to 24 h.

Table 8a Typical Formula of Liquid Three-Stage Sourdough: Stage 1—Flour Cooking

	Typical value	%
	301 kg	100.0
Rye flour	80 kg	27.0
Water (+90 °C)	200 kg	66.0
Rye malt	20 kg	6.6
Caraway seeds	1.4 kg	0.5
Cooking time	60 min	
Cooking temp.	80 °C	
Cooked flour yield	300	

Table 8b Typical Formula of Liquid Three-Stage Sourdough: Stage 2—First Fermentation

Factor	Typical value	%
	248 kg	100.0
Cooked flour (+50 °C)	165 kg	67.0
Starter	83 kg	33.0
Fermentation time	2 h	(typical variation) (2–3)
Fermentation temp.	50 °C	(45–55 °C)
Total titratable acids of ripened sourdough	9–12	

Table 8c Typical Formula of Liquid Three-Stage Sourdough: Stage 3—Second Fermentation

Factor	Typical value	%
	248 kg	100.0
Starter	124 kg	50.0
First ferment	124 kg	50.0
Fermentation time	3 h	(2.5–3.5)
Fermentation temp.	30 °C	(27–33)
Total titratable acids of ripened sourdough	16–18	

Table 9 Specification of R1800 Rye Flour, R1450 Rye Flour, and R750 Rye Flour for Sourdough

		Flour type	
Measurement	R1800	R1450	R750[a]
Ash content	1.74 ± 0.05% d.m.	1.42 ± 0.03% d.m.	0.74 ± 0.03% d.m.
Falling number	110 ± 20	120 ± 20	130 ± 20
Moisture content	13.3 ± 0.5%	13.0 ± 0.5%	13.3 ± 0.5%

[a] Seldom used in sourdough.
Source: Rigas Dzirnavnieks, Latvia.

After reaching a temperature of + 50 °C the cooked flour is fermented by mesophilic lactobacilli (*Lb. plantarum, Lb. casei, Lb. brevis*). This first fermentation is carried out typically in 2 h ending with a total titratable acidity of 9–12. The recommended temperature is 45–55 °C. The amount of starter is relatively high, approximately one-third (33%) of the final mass. After fermentation, two-thirds of the sour is pumped to subsequent steps of the process while one-third serves as starter for the next batch. After mesophilic fermentation, the temperature of the sourdough is lowered to + 30 °C (+27–33 °C).

The aerobic sour contains pure cultures of Lactobacilleae (*Lb. fermenti*) and yeast (*S. cerevisiae*). Sour from mesophilic fermentation is added to the aerobic sour at a ratio of 1:1. Using intensive agitation the final sourdough is ripened in 3 (2.5–3.5) h time. This is the final sourdough, which is then used in dough. Fifty percent of the final sourdough is used as a starter for a new sourdough batch. The amount started is very high, thus ensuring sufficient gassing power (yeast) in sourdough. A baker's yeast addition in bread dough is normally not needed because the yeasts present in sourdough have enough gassing power for a normal dough leavening.

Stiff Sourdough. Stiff sourdough is usually made in dough mixing bowls, where the final dough is mixed after the sourdough ripening period. This method is often used in small bakeries or if a liquid system is not available. The flour/water ratio is 60:40, the ratio of ordinary rye dough (Table 6). The consistency of stiff sourdough can be compared to that of normal rye dough.

The amount of starter may vary from 25% to 50% of the amount of final sourdough. The more starter is used the shorter the ripening time. The flour type affects the acidifying properties; after 3.5 h fermentation time total titratable acidity is 15 with R1450 (ash content 1.42% d.m.) and 13 with R1800 (ash content up to 1.9% d.m.). This type of sourdough allows large variation in fermentation conditions without changing the product quality remarkably. The pure cultures used for stiff sourdough are *Lb. plantarum, Lb. brevis,* and *Lb. brevis*; yeast is typically *S. minor* [40].

D. Products

Traditional rye breads in the Baltic countries mostly are oval in shape; the weight of the loaf may vary from 0.5 to 5.0 kg. The most common weight is 0.7–1.0 kg, in Lithuania commonly even 2.0 kg. There are some round bread varieties also. Due to the flour cooking and gelatinizing processes, bread is very moist and chewy and has a long shelf life.

The acidity of rye bread is most commonly 6–8 TTA, very seldom more than 10. There are no data available of the acid composition of Baltic rye bread. Due to the relatively low acidity and high content of syrup (or sugar), Baltic rye bread is always more sweet than acidic in taste.

In the 1990s all of the bread produced was unpacked. Today practically all of the bread is packaged in Lithuania, more than half in Estonia and nearly half in Latvia. Traditional bread is still a very important part of the total bread market. Central European and Scandinavian influence is bringing new ideas to the Baltic bread market, but the strong tradition will remain.

REFERENCES

1. Rukiin viljely (Cultivation of Rye), Maatalouskeskusten liiton julkaisuja No. 682 (Publications of Rural Advisory Organization) (M. Kinnanen, ed.), 1983, p. 4 (in Finnish).
2. AC Nielsen ScanTrak, The Sales of EAN-coded Products in Finland (2000).

3. Lahti-Koski, M. and Kilkkinen, A. Ravitsemuskertomus 2000 (Nutrition Report 2000), Hakapaino Oy, Helsinki, Finland, 2001, p. 12.
4. Anttolainen, M., Javanainen, J., Kaartinen, P., Lahti-Koski, M., Lauronen, J., Männistö, S., Ovaskainen, M.-L., Paajanen, H., Pietinen, P., Roos, E., Valsta, L. and Virtanen, M. The 1997 Dietary Survey of Finnish Adults, National Public Health Institute, Helsinki, Finland, 1998, p. 52.
5. Adlercreutz, H. Environ. Health Persp. *103*:103 (1995).
6. Borriello, S. P., Setchell, K. D. R., Axelson, M. and Lawson, A. M. J. Appl. Bactial, *58*:37 (1985).
7. Heinonen, S-M., Nurmi, T. and Adlercreutz, H. Whole Grain and Human Health — International Symposium (K. Liukkonen, A. Kuokka, and K. Poutanen, eds.). Porvoo, Finland, 2001, p. 80.
8. Consensus Statement on Cereals, Fibre and Colorectal and Breast Cancers. ECP Consensus Panel on Cereals and Cancer. Eur J Cancer Prev. 7(Suppl.2):1 (1998).
9. Pietinen, P., Rimm, E. R., Korhonen, P., Hartman, A. M., Willett, W. C., Albanes, D. and Virtamo, J. AHA Circ. *94*:2710 (1996).
10. Salovaara, H. Paluu juuriin—hapanraskin käyttö leipomoissa 1 (Back to the roots — The use of sourdough in bakeries). Leipuri (9):32 (1993) (in Finnish).
11. Hurri-Martikainen, M. Bakery specific sourdough starter. Masters thesis, In Finnish. University of Helsinki, Department of Food Technology. EKT Series 787 (1989).
12. Itkonen, J. V. Leipurin kirja (Baker's Book), Otava, Helsinki, Finland, 1976, p. 79 (in Finnish).
13. Grönman, S. and Möttönen, K. Production of acids in rye flour baking. Proc. 7th World Cereal and Bread Congress, Prague, 1982.
14. Salovaara, H. Lactic Acid Bacteria—Technology and Health Effects (S. Salminen and A. von Wright, eds.). Marcel Dekker, New York, 1998, p. 115.
15. Salovaara, H. International Rye Symposium: Technology and Products (K. Poutanen and K. Autio, eds.). VTT Symposium 161: Helsinki, Finland, 1995, p. 188.
16. Korhonen, R. Herstellung von finnischen Fladenfrischbrot (Production of finnish flat bread). Getreide Mehl Brot, *85*:85 (1985).
17. Korkman, M. Tarinaa näkkileivästä (Story of crispbread). Kansanterveys *12*:35 (1970) (in Finnish).
18. S. Bressler, *Food Engineering Int'l* 34–35 (1979).
19. Eliasson, A.-C. and Larsson, K. Cereals in Breadmaking: A Molecular Colloidal Approach. Marcel Dekker, New York, 1993.
20. Autio, K., Koskinen, M. and Parkkonen, T. International Rye Symposium: Technology and Products (K. Poutanen and K. Autio, eds.). VTT Symposium 161: Helsinki, Finland, 1995.
21. Hoseney, R. C. Food Technol. *1*:114 (1984).
22. Autio, K. and Salmenkallio-Marttila, M. Lebensm. Wissen. Technol. *34*:18–22 (2001).

23. Autio, K. Carbohydrates in Food (A.-C. Eliasson, ed.). Marcel Dekker, New York, 1995, p. 227.
24. Stear, C. A. Handbook of Breadmaking Technology. Elsevier Science, Amsterdam, 1990.
25. Lönner, C. and Ahme, S. Food Biotechnology Microorganisms (Y. Hui and G. Khacharourians, eds.). VCH, New York, 1995.
26. Hammes, W. P. and Gänzle, M. G. Microbiology of Fermented Foods (B. J. Wood, ed.). Blackie, London, 1998, p. 199.
27. Salovaara, H., Sareneva, T., Savola, P., Kojo, K., Väisänen, E. and Sinda, E. Yeast Strains Isolated from Sour Rye Doughs: Screening of Technological Potential in Baking. Nordisk Industrifond, Project P656BU7/85" Frysning af gaerdehe–bedre brod. Teknisk rapport (1989).
28. Ollila, U. Hapanleivonnan mikrobien valintaperusteet (Selection criteria for sourdough microbes). Helsingin yliopisto, mikrobiologian laitos. Masters thesis. In Finnish. University of Helsinki, Department of Microbiology, 1991.
29. Hansen, Å. 24. Nordic Cereal Congress, Stockholm, Sweden, 1992.
30. Salovaara, H. and Valjakka, T. Int. J. Food Sci. Technol. *22*:591 (1987).
31. Hansen, Å. and Lund, B. Lebensm. Wissen. Technol. *22*:141–144 (1989).
32. Spicher, G. and Stephan, H. Handbuch Sauerteig. Biologie, Biochemie, Technologie, B. Behr's Verlag GmbH, Hamburg, 1993.
33. Spicher, G., Schroeder, R. and Stephan, H. Z. Lebensm. Unters. Forsch. *171*:119 (1980).
34. Helleman, U., Tuorila, H., Salovaara, H. and Tarkkonen, L. International J. Food Sci. Technol. *22*:693 (1987).
35. Hellemann, U., Powers, J. J., Salovaara, H., Shinholser, K. and Ellilä, M. J. Sensory Studi. *3*:95 (1988).
36. Hellemann, U. J. Food Sci. Technol. *27*:201 (1992).
37. Heiniö, R.-L., Urala, N., Vainionpää, J., Poutanen, K. and Tuorila, H. Int. J. Food Sci. Technol. *32*:169–178 (1997).
38. Katina, K., Sauri, M., Alakomi, H.-L. and Mattila-Sandholm, T. Lebensm. Wissen. Technol. (accepted).
39. Rosenquist, H. and Hansen, Å. J. Appli. Microbiol. *85*:621 (1998).
40. *Sbornik technologiceskih instrukcij dija proizwodstwa hlebobulocnih izdelij.* (Handbook of Technological Instructions for Bread Production), Soviet Ministry of Bread Products. Moscow, 1989, p. 60 (in Russian).
41. Salovaara, H. and Katunpää, H. Acta Aliment. Poloni. *10*:231 (1984).
42. Salovaara, H. and Savolainen, J. Acta Aliment. Poloni. *10*:241 (1984).
43. Haapanen, J. Baking technology of rye dough and weekend maintenance of seed sour. Masters thesis. In Finnish. University of Helsinki, Department of Food Technology. EKT Series 1139 (1998).
44. Joutsen, M. The microbiology of cold stored sourdough. Master's thesis. In Finnish. University of Helsinki, Department of Food Technology. EKT Series 1071 (1997).
45. Nieminen, A., Lilja, M., Levola, M. and Salovaara, H. International Rye Symposium: Technology and Products (K. Poutanen and K. Autio, eds.). VTT Symposium 161: Helsinki, Finland, 1995.

Index